ISBN: 978-1-387-93181-1 (Paperback)
ISBN: 978-1-387-93181-1 (Hardcover)

Issued and Referenced under Library of Congress Control

Any references to historical events, real people, or real places are used fictitiously. Names, characters, and places are products of the author's imagination.

Front cover image by Lulu Press, Inc.
Book design by Lulu Press, Inc.

Printed by Lulu Press, Inc., in the United States of America.

First printing edition 2018.

Lulu Press Inc.
Morrisville, North Carolina, United States of America

ACKNOWLEDGEMENT

This work would have not been possible without the blessings of almighty. I am highly grateful to God for his constant grace and divine sanction. The success and final outcome of this project required a lot of guidance and assistance from many people and I am extremely privileged to have got all this all along the completion of my project. In the due course of this work, my family has supported me a lot. I would like to thank my parents Dr. Gopal Sachdev and Neena Sachdev, my brother Sidharth Sachdev and sister in law Suvidha Chopra, whose love, support and guidance are with me in whatever I pursue. They are the ultimate role models which I have in my life.

I am thankful and fortunate enough to get constant support and guidance from my teachers and friends which helped me in successfully completing this work.

Dr. Rohan Sachdev

INTRODUCTION

Sucrose (table sugar) is considered as sugar by most of the people. The term sugars include all the monosaccharides and disaccharides, the most common of which are glucose, fructose, sucrose, maltose and lactose. Though sweet, these sugars are the causes for many bitter experiences faced by the modern day civilized man. Sugar, directly or indirectly is considered as culprit for many diseases like diabetes, obesity, atherosclerosis, etc., which prompted the search for a suitable substitute.' At this given point of time, there is no such substitute which can replace sugar (which is versatile) in all aspects.

The dental profession shares an interest in the search for safe, palatable sugar substitutes, as there is established evidence suggesting the causal relationship between sugar and dental caries. Dentistry has evolved through years from conventional "drilling and filling" stage to "preventive" stage following the famous saying "Prevention is Better than Cure". Use of sugar substitutes in preventive dentistry is gaining importance. Replacing sugar with a suitable sugar substitute to combat dental caries is an option wide open as significantly better dental properties were observed when compared to sucrose and glucose. Chewing xylitol gum caused significantly lower net progression of decay and had inhibitory effect on mutans streptococci in saliva and dental plaque and also on lactobacilli in saliva was observed. Acceptance of new products by people is most important and difficult aspect in their success. Xylitol chewing gum became a well adopted practice among Finnish adolescents, which is an example of the positive effect of health education given by comprehensive, preventively oriented system of dental health care in association with commercial interests!) Studies were done to find the difference between sugar substitutes and other preventive measures like pit and fissure sealants, no statistically significant difference in caries increment was observed between sealant and xylitol groups. 10% xylitol, when added to a triclosan containing, dentifrice reduced the number of MS in saliva and dental plaque. Acid production by bacteria, which is the main cause for dental caries was tahibited by xylitol. Sugar substitutes like xylitol when combined with other compounds like fluoride showed synergistic effect in inhibiting the acid production by mutans streptococcus (MS). Remineralization potential of xylitol chewing gum when compared to mastic chewing gum was attributed to increasedsalivation," but mean degree of remineralization was greater when combined with calcium lactates- or funoran and calcium hydrogen phosphate.Antimicrobial activity of stevioside against periodontal pathogens like PorphyTomonas gingivalis and Aggregatibacter Actinomyeetem comitans was significant. It also showed antifungal activity against candida albicans.

The words, "sugar-free", "zero-calorie sugars" or "calorie-free" are commonly heard or seen on advertisement hoardings now-a-days. A food may have the words 'sugar-free' on the front label, but that does not mean the food is carbohydrate-free or calorie-free. The use of these zero-calorie sugars is increasing, as a result of increased body physique consciousness and weight reduction programs, rather than dental health. Instead of attractive and influencing advertisements given by manufacturers on sugar substitutes, many confusions and myths prevail in the public regarding them. This confusion may be from the unfaithful information available on the internet. Here is a small effort to give an elaborate discussion on sugar substitutes and their role in dental health and also remove the myths about them and give a clear cut idea on them.

REVIEW OF LITERATURE

Jensen ME (1986)[15]conducted a study to examine the interproximal plaque pH responses a five common acidogenic snack foods and also the effect of chewing sorbitol sweetened gum after the consumption of these snack foods. A wire telemetric appliance containing a pH microelectrode was used for obtaining interproximal plaque pH data from five volunteers who =ceded complete crown restorations on mandibular permanent molars. Ten test sessions were conducted for each of the five volunteers in two sets. The authors observed rapid decrease in plaque pH for extended periods. In a second set of sessions, volunteers chewed sugarless gum for 10 minutes, starting 15 minutes after they ate the snack food. Increase in plaque pH was observed, which remained at a level considered safe for teeth for 30 minutes after chewing the

Isokangas et al. (1989)[3]conducted a study to analyze the increment of dental caries during three post-experimental years(1984-1987) in subjects who used xylitol chewing gum for 2years, and during two post-use years (1985-1987) in high risk subjects who used xylitol for 3years. It was a blind study in which re-examinations were carried out by two dentists who had participated in the original studies (Isokangas, 1987). One hundred and fourty seven children of the original xylitol group and 122 children of the original control group were clinically re-

6

examined in 1987, i.e., 3years after the discontinuation of the daily use of xylitol gum in the main experimental group (2years after discontinuation in the high risk groups). The authors observed that caries reduction was greater during the post-experiment years 1984-87 than during the actual field trial in 1982-84. By the results, it was suggested that the value of xylitol in caries prevention depends on the timing of the treatment in relation to the development of the dentition.

Grenby et al. (1989)[16] studied the dental properties of lactitol compared with five other bulk sweeteners in-vitro. The sweeteners tested were lactitol monohydrate in the form of a coarse white powder and pure grades of glucose, sucrose, sorbitol, mannitol, and xylitol as either fine crystals or powders. Standardized mixed cultures of dental plaque microorganisms were incubatedfor 24 hours in media containing six different bulk sweeteners as energy source. The authors observed that demineralization was most severe with glucose and sucrose. Less acid was zenerated from sorbitol and mannitol, with much reduced demineralization.Fermentation of -ctitol and xylitol was only very slight. Rapid multiplication of bacteria was observed in the gag& media and least of all in the lactitol and xylitol media.Significantly less polysaccharide was synthesized from sorbitol and mannitol. Lactitol and xylitol followed, with xylitol giving a significantly lower figure than all the others. The reductions in calcium and phosphorous ssolving in the polyol media compared with the glucose media were:

mannitol 63-69%, sorbitol approximately 80-85% and lactitol and xylitol 94-98%

Kandelman and Gagnon (1990)[4]conducted a study to test the effect of chewing gum :ontaining xylitol on the incidence and progression of dental caries through a 24 month field trial :onducted in Montreal. The subjects were divided into two experimental groups (15 percent and 65 percent xylitol chewing gum distributed three times a day at school) and one control group (without chewing gum) and were exposed to the same basic preventive program. Significantly lower net progression of decay(progressions-reversals) was observed in the experimental group. Chewing xylitol gum had beneficial effect on the caries process for all types of tooth surfaces, and especially for bucco-lingual surfaces.DMF(S) increment of 2.24 surfaces, compared with 6.06 surfaces for the control group was observed. Results for the plaque index were in agreement with those of the DMF(S) increment and the net progression of decay. The authors concluded that the additional use of xylitol containing chewing gum in a school preventive program have an impressive reduction in caries incidence.

Dawes and Macpherson (1992)[17]conducted a study to observe how salivary flow rate and pH vary with time during use of chewing gums and lozenges. One unflavored gum, six flavored chewing gums and two lozenges were employed. Gum one was about 1.0gm gum base, gums 6 and 7 and lozenges 8 contained sucrose, while other products were sucrose-free. Gum4 contained 1.5% of total organic acids. Participants collected unstimulated saliva and then, on different occasions, chewed one of six flavored gums, or gum base, or sucked on one of two lozenges for 20min, during which time eight separate saliva samples were collected over the timeperiods 0-1,1-2,2-4,4-6,6-8,8-10,10-15 and 15-20minutes.The authors observed that flow rate peaked duringfirst minute of stimulation with all nine products with lozenges, flow rate fell towards the unstimulated rate when the lozenges had dissolved.The authors concluded that all the tested chewing-gums and lozenges stimulated salivary flow initially and fell with continued chewing.

Wennerholm et al. (1994)[5] have studied the effect of xylitol -and sorbitol in chewing gums on Mutans Streptococci, plaque Ph and mineral loss of enamel. It was a randomized, cross-overstudy. Twenty subjects having more than 3x105mutans streptococci per milliliter of saliva during screening examination were selected for the study. Four different chewing gums, containing: (1)70% xylitol, (2)35% xylitol+35% sorbitol, (3)17.5% xylitol+52.5% sorbitol, and (4)70% sorbitol, were tested. The participants used 12 pieces of each gum per day for 25 days. During the four experimental periods, they wore a removable palatine plate containing two demineralized enamel samples and were asked to use non-fluoridated tooth paste. Saliva and plaque samples were collected before and after

9

the chewing-gum periods. The authors observed that with an increased concentration of xylitol in the gum resulted in a lower number of mutans streptococci in both saliva and dental plaque, although the decreases were only significant in the saliva samples($p<0.01$). By this study, the authors concluded that xylitol has an inhibitory effect on mutans streptococci in saliva and dental plaque and on lactobacilli in saliva.

Honkala et al. (1996)[6] studied the adoption of xylitol chewing gum in Finland using data from two comparable postal surveys for national samples of 12-18 year olds in 1977 and 1991. Only 12% of this age group used xylitol chewing gum in 1977 which had become common in 1991 (64% of boys, 81% of girls). Since 1977 the adolescent health and lifestyle survey data have been collected biannually by mailed questionnaires. Response rate was lower among boys than among girls in both the years. The authors observed that the proportion of chewing gum increased from 90 to 94% among boys and from 89 to 97% among girls between 1977 and 1991 and xylitol became the most common choice. Daily use of xylitol chewing gum did not vary according to socioeconomic status or level of urbanization. The increase in use of xylitol gum is an example of the positive effect of health education given by comprehensive, :eventively oriented, system of dental health care in association with commercial interests.

Scheie et al. (1998)[18] tested the hypothesis that chewing of xylitol or xylitol/sorbitol-containing chewing gum reduces plaque formation and the acidogenic potential of dental plaque. Thirty healthy volunteers aged from 19 to 28 years were randomly allocated into xylitol, xylitol/sorbitol, or sucrose —sweetened gum chewing gum groups. A three day plaque accumulation period of no oral hygiene was instituted prior to and at the termination of the 33 day program. Plaque quantity was assessed on the basis of protein content of plaque samples and the acidogenic potential by the quantity of various organic acids formed from D-glucose. The authors observed that plaque formation, acidogenic potential and glycolytic profiles were similar at baseline and after the gum-chewing periods. Also, no intracellular accumulation of glycolytic metabolites was observed. The authors concluded that xylitol or xylitol and sorbitol has no effect on the microbial deposits on the teeth of young adults with low caries experience.

Makinen et al. (1998)[19] investigated the effect of usage of xylitol(X) — Malitol-syrup (MS) and xylitol (X) — polydextrose (PD) saliva stimulated on some biochemical properties of human whole saliva through a 4-month pilot study. One hundred and eighty— eight young subjects (mean age, 22years) of both sexes (75% females) were randomly divided into three groups of equal size. The subjects in one group used xylitol (X) — maltitol syrup (MS), while another group received xylitol (X) — polydextrose (PD). Subjects in the third (comparison) group did not

11

receive saliva stimulants. Paraffin stimulated whole saliva samples were collected at baseline, after 2 months and 4 months for microbiologic and biochemical tests. The usage of X-MS was associated with a significant ($P < 0.05$) reduction in the salivary sucrase activity. After 4 months, the activity of enzymes hydrolyzing N(alpha)-benzoyl-DL-arginyl-p-nitroaniline was significantly reduced in all groups, while the levels of free sialic acid were reduced in group X-PD only ($P < 0.05$). These salivary changes most likely reflected microbial shifts in the oral cavity and suggest that information from saliva studies may be of avail when deciding which bulking, agents should be used in xylitol-based saliva stimulants.

Simons et al. (1999)[20]determined the clinical effectiveness of chewing gums containing xylitol (X) or a combination of xylitol and chlorhexidine (CHX). It was a double-blind,randomized, cross over, 5-day clinical trial, with a 9-day washout period in a group of participants over 40 years old. The investigator (DS) collected clinical data, and verified intra and inter-examiner diagnostic consistency. Another dentist (FC) re-examined 6 subjects at the baseline and second stage examination. After professional tooth cleaning, eight subjects used two pieces of a chlorhexidine acetate/ xylitol gum (ACHX -a liquor ice flavored CHX/X) gum, 2 pieces of BCHX (a chocolate mint flavored CHX/X), 2 pieces of X (a liquor ice flavored X gum) and 1 piece of ACHX in a random order. Gums were chewed two times daily for 15 minutes and volunteers refrained from all other oral hygiene procedures. It was observed that both ACHX and BCHX gums demonstrated significantly lower plaque levels compared to the X gum. By theresults, theauthors concluded that regular use of chlorhexidine acetate (CHX) containing gum can control dental plaque formation and reduce gingival inflammation.

Giersten et al. (1999)[21] tested the hypothesis that xylitol, alone and in combination with fluoride, affects the salivary flow rate and micro-biota, dental plaque accumulation, gingivitis development and the acidogenic potential of plaque. In a double-blind controlled trial, three groups, each of 10 subjects, rinsed for 1 minute,three times daily over two 4-week periods, first with 10ml water (control), and thereafter with either 0.05%NaF, 40% xylitol, or with 0.025%NaF plus 20%xylitol. Habitual mechanical cleaning was performed during the first 2weeks of each period but abstained from interdental cleaning during the final 2 weeks. Oral hygiene was discontinued for the last 2 days of each period to permit plaque accumulation. The following parameters were assessed: 1) unstimulated and paraffin-stimulated salivary secretion rates; 2) salivary microbiota; 3) plaque index; 4) papillar bleeding; 5) plaque pH response to sucrose, and 6) lactate formation by dental plaque. The authors observed no statistically significant difference in any of the parameters. The authors concluded that, three daily mouth rinses with xylitol and fluoride, separately or in combination, did not affect the salivary flow rate or micro-biota, dental plaque accumulation, gingivitis development, or the acidogenic potential of plaque.

Hujoel et al. (1999)[22] conducted a cohort study in 510 children to report the optimum -ime to initiate habitual xylitol gum-chewing for obtaining long term caries prevention. The goal was (1) to determine if sorbitol and sorbitol/xylitol mixtures provide a long-term benefit, and (2) :o determine which teeth benefit most from two-year habitual gum-chewing those erupting before, during, or after habitual gum chewing. Children, on average six years old, chewing gums sweetened with xylitol, sorbitol, or xylitol/sorbitol mixtures. There was no-gum control group. Five years after the two-year program of habitual gum chewing ended, 288 children were re-examined. The authors observed that sorbitol gums had no significant long-term effect whereas xylitol gum and, to a lesser extent, xylitol/sorbitol gum had a long-term preventive effect. The Iong-term caries risk reduction associated with xylitol strongly depended on when teeth erupted. No significant long-term prevention was observed on the teeth that erupted before the start of gum-chewing. The authors concluded that habitual xylitol gum-chewing should be started at least one year before permanent teeth erupt for maximum caries-preventive effects.

Honkala et al. (1999)[23] had studied to find out how common the recommended habit of using xylitol chewing gum on a daily basis was among Finnish schoolchildren, which was a part of comprehensive cross-national survey on Health behavior in school-aged children (HBSC Study)- a World health organization collaborative study. The data were collected using standardized questionnaires to which pupils in grades 5-11 years, 7-13 years, and 9-15 years responded anonymously in school classrooms. It was observed that response rate varied between 87% (15-year-old boys) and 94% (11- and 13-year old girls). The percentages of daily users of xylitol gums among boys were 17% (11 years), 46% (13 years), and 44% (15 years), and among girls it is 57% (11 years), 65% (13 years), and 69% (15 years), respectively. It was concluded that since 1991 the use of xylitol chewing gum has further increased in Finland and currently more than a half of all schoolchildren benefit from it.

Lam et al. (2000)[24] had studied to test the acceptability of xylitol-based snack foods in young children. Xylitol containing snack foods were developed and tested in a convenience sample of 31 children aged 3 to 6 years. In the first phase, each child was a presented with a tray of six Xylitol-based foods (popsicles, pudding, gumdrops, gelatin dessert, cookies, and popcorn) and asked to taste each item in any desired order. Immediately after tasting a food, the child was iskedto place it in front of one of three cartoon faces (smile, frown, neutral) representing the response to the taste of that particular food. In the second phase, the child was asked to rank order the foods in each face category (smile, frown, neutral). Ranks within categories were then combined to obtain a rank ordering for all the foods. The authors observed that pudding was significantly less preferred than the other foods. At least 84% of the children found five of the six foods very good or satisfactory, when considered individually. These results suggest that xylitol containing snack foods are generally well accepted by children.

Alanen et al. (2000)[25] had studied to test the caries-preventive effect of xylitol candies in school children with erupting permanent teeth. A 5-year follow-up study with 2- or 3- year xylitol consumption periods began in Estonia in 1994 with 740,ten year old children in twelve schools at baseline examinations. Based on baseline caries experience schools were divided into three clusters each including three to five schools. The school classes were randomly assigned within these areas to control, xylitol chewing gum and xylitol candy groups. The products were used threetimes per day under teacher's supervision. The daily dose of xylitol was 5g in all groups. The children were examined every year in September by two experienced clinicians. After three years, the authors observed highly significant 35%-60% reduction in caries incidence in all xylitol groups compared to control groups. This study suggests that not only xylitol chewing but also xylitol candies are effective in caries prevention, and that a school based delivery system seem to offer a practical way to distribute and control the use of the xylitol products.

Alanen et al. (2000)[7] had compared the regular use of xylitol chewing gum during 2 or 3 school years with application of occlusal sealants in a randomized study. Fifth grade children in fourteen schools in Hameenlinna and the surrounding four communities together forming the Hameenlinna health care center, were invited to participate in a five year trial. An informed :onsent form was sent to all children and their parents at their home addresses. The fourteen .7a.rticipating school classes were randomly assigned as clusters into the sealant, the two year -Oho' chewing gum, and the three year xylitol chewing gum groups. The dentists treating the :hildren were asked to apply sealants on an individual basis when indicated for the children in he sealant group, butnever in the xylitol groups. The recommended chewing time for the xylitol zroup was about ten minutes. The authors observed no statistically significant difference in varies increment figures between sealant and xylitol groups. The authors concluded from their results suggesting that the selection between the sealant and xylitol should be based on practical aspects, such as cost of treatment, occurrence of caries, co-operation between school and healthcare, availability of healthcare personnel and equipment opportunity costs etc.

Jannesson et al. (2002)[8] had evaluated the effect of the combination of triclosan and xylitol in tooth paste on mutans streptococci(MS) in saliva and dental plaque after six months use. It was a double-blind clinical study done in 155 individuals with >105MS/m1 saliva, dividing into three groups (n=51-52) balanced according to their MS counts at baseline. Whole saliva and pooled plaque samples were obtained after 2, 4 and 6 months. The authors observed that total-xylitol group showed significant reduction (p<0.001) in MS counts within the groups for saliva and plaque samples at all three occasions. The authors concluded that addition of 10% xylitol to a triclosan-containing dentrifrice reduces the number of MS in saliva and dental plaque.

Roberts et al. (2002)[26] had examined the effect of xylitol on levels of Streptococcus mutans and S.Sobrinus and its mechanism of action. They compared cariogenic bacteria levels before and after exposure to xylitol products in 187 children and two adults respectively. Unstimulated saliva samples were taken from all subjects, in addition they collected gingival samples from adults. Samples were plated on microbiological media with varying concentrations of xylitol. Specific DNA probes were used for identification and Pulse field gel electrophoresis for genetic relatedness. They observed that S.mutans levels remained stable before and after xylitol exposure and also increased tolerance to xylitol.

Assev et al. (2002)[27] had compared cariogenic traits in xylitol resistant (X-R) and xylitol-sensitive (X-S) strains. Six strains of Mutans Streptococci, three X-R and three X-S strains, wereudied. Xylitol resistance and sensitivity were confirmed by growth in xylitol-supplemented media. Acid production was initiated by adding it-

18

labelled glucose, fructose or xylitol to bacterial suspensions. The authors observed lactate as the major metabolite from glucose in the presence or absence of xylitol. Lactate production per colony forming unit was lower in X-S cells than in X-R cells. Both X-R and X-S cells metabolized fructose. Both cells took up xylitol, but xylitol-5-P was detected in X-S cells only. The authors found no difference in polysaccharide production. This study didn't support the contention that X-R are less cariogenic than X-S mutans streptococci.

Miyasawa et al. (2003)[9] had studied the inhibitory effect of xylitol on the acid production of S.mutans at several pH levels under strictly anaerobic conditions found in the deep layer of dental plaque.Xylitol inhibited the rate of acid production from glucose and changed the profile of acidic end products to formate-acetate dominance, with a decrease in the intracellular level of fructose 1,6-bisphosphate and an intracellular accumulation of xylitol 5-phosphate (X5P). These results were notable at pH 5.5-7.0, but were not evident at pH 5.0. Since the activity of phosphoenol-pyruvate phosphor-transferase for xylitol was greater at higher pH, it is suggested that xylitol could be incorporated more efficiently at higher pH and that the resultant accumulation of X5P could inhibit the glycolysis of S. mutans more effectively.

Blicks et al.(2004)[28] had studied to investigate the effect of two contrasting doses pf xylitol on Mutans Streptococcus(MS) counts in saliva and plaque collected adjacent to fixed orthodontic appliances and to investigate the relationship between xylitol-susceptible(Xs) and xylitol resistance(XR) strains in saliva, with a null hypothesis assumpting no changes. Fifty six randomly selected adolescents and young adults (mean age 15.8 yrs.) of both sexes were selected after taking consent from their carers. They were assigned into following groups: A,(n=23) two xylitol tablets two times a day(1.7g xylitol/d) for 18 weeks.; B,(n=23) two tablets four times per day(3.4g xylitol/d) for 18 weeks.; and C,(n=10) no tablets. Stimulated saliva and plaque samples vie collected at baseline and after 6, 12 and 18weeks. By the results, it was concluded that labitual intake of xylitol containing tablets could affect the bacterial composition in saliva, but long term alteration in MS counts or lactic acid formation rate.

Makinen et al. (2005)[10] had studied to investigate the effects of 6-month use of rythritol, xylitol, and D-glucitol on the presence of mutans streptococci in whole saliva and dental plaque in high school-age subjects. One hundred and thirty six participants from 152 baseline sample were divided into four test groups (erythritol, xylitol, D-glucitol and control zroup). It was observed use of erythritol and xylitol was associated with statistically significant reduction in the plaque and saliva levels of mutans streptococci. Amount of dental plaque was also significantly reduced in subjects receiving erythritol and xylitol. Although the biochemical mechanism differs, erythritol and xylitol may exert similar effects on some risk factors of dental caries.

Holgerson et al. (2005)[29] had studied to measure the xylitol concentrations in saliva after use of selected xylitol containing products and to investigate whether or not a dose relationship would be obtained in plaque. The study consists of two sets of experiments one in saliva (single-blind design) and one in plaque (single-blind crossover design). Twelve children (6 boys, 6 girls, mean age 11.5 years) were included in the study after taking informed consent from their parents. Various xylitol products : (A) chewing gums (1.3g xylitol), (B) sucking tablets (0.8g xylitol), (C) candy tablets (1.1g xylitol), (D) tooth paste (0.1g xylitol), (E) rinse (01.0g xylitol), and (F) a non-xylitol paraffin were used in the study. Unstimulated saliva was sampled 1, 3,8,16 and 30 min after use. The xylitol concentration in the saliva was determined enzymatically. The authors had observed that statistically significant elevations of salivary xylitol levels were demonstrated for all products during the first 8-16 min when compared with baseline ($P<0.05$). The highest mean value in saliva was obtained after use of chewing gums (A) and lowest was demonstrated after using tooth paste (D). In dental plaque, a statistically significant increase was observed. The authors concluded that xylitol containing products gave significantly elevated concentrations of xylitol in whole saliva for 8-16 min and a tendency for a dose-response relationship was disclosed in dental plaque.

Maeharaet al. (2005)[30] had conducted a study to evaluate the combined inhibitory effect of fluoride and xylitol on the acid production of mutans streptococci from glucose under strictly anaerobic conditions at fixed pH 5.5 and 7.0. The bacteria were grown in a tryptone yeast extract. Acidic end products of glucose fermentation and intracellular glycolytic intermediates were assayed. The authors observed that the combination of fluoride and xylitol have synergistic effect in inhibiting acid production. The proportion of lactic acid in the total amount of acidic endproducts decreased, while the proportion of formic and acetic acids increased in their combination. Analyses of intracellular glycolytic intermediates revealed that xylitol inhibited the upper part of the glycolytic pathway, while fluoride inhibited the lower part. The authors concluded that synergistic inhibition by combination of fluoride and xylitol on glycolysis by mutans streptococci is due to direct inhibition of glycolytic enzymes. The study suggests that xylitol has the potential to enhance inhibitory effects of low concentrations of fluoride.

Suda et al. (2005)[12] had conducted a study to determine whether the effect of remineralization of enamel surface by xylitol can be enhanced when calcium lactate is added to gum containing xylitol. Ten volunteers (2males and 8 females, age 22-27yrs) who met the requirements were recruited after taking informed consent. Enamel slabs were cut from premolars and impacted third molars with sound enamel surfaces. The slabs were immersed in 40 ml of demineralization buffer consisting of 20g/l carbopol 907(carboxy-polymethylene), 500mg/lit hydroxyl apatite and 0.1 M lactic acid ,ph4.8,for 4days at 37°C which was changed after 2days after demineralization, the slabs were cut into two halves of which one is used as control and stored in milk water. These slabs were mounted to an intraoral acrylic palatal appliances. The authors observed that mean degree of remineralization was greater after chewing xylitol-calcium gum. Finally authors concluded that chewing gum containing xylitol+calcium lactate enhance remineralization.

Honkala et al. (2006)[31] had done a field study to test the efficacy of xylitol candies in renting caries among individuals in two special schools on Kuwait. Two schools (one for xysand one for girls) for physically disabled individuals which had shown a high caries experience among all schools for students of special needs were selected. Only caries extending aato dentine was recorded as in the W.H.O criteria. School health nurses distributed xylitol zandies to the students 3 times during the school day (after breakfast and lunch, and before -leaving the school).The authors observed that in the xylitol group, the baseline DS and DMFS mores were 3.4 and 8.2 and in the follow-up 1.9 and 7.1, respectively whereas in the control the baseline scores were DS 3.9 and DMFS9.8, and the follow-up scores DS 3.9 and

DMFS13.2. The authors concluded that xylitol have a strong preventive and a clear :mineralizing effect on caries.

Holgerson et al. (2006)[32] had investigated the effect of xylitol on mutans streptococci and lactic acid formation in saliva and the amount of visible dental plaque after a 4-week period. They also explored the possible differences between children with and without caries experience with a null hypothesis assumpting no difference. It was a randomized double-blind prospective design with two parallel arms. One hundred and fourty nine pupils in grades 1-6 in a comprehensive school in northern Sweden were invited of which 128 children consented to participate. The children were stratified as having caries experience (DMFS/dmfs>=1) or not before the random allocation to a test or control group. The control group(A) was given two pellets, containing sorbitol and maltitol three times daily for 4 weeks, and the test group(B) received corresponding pellets with xylitol as single sweetener(total dose=6.18g/day). The outcome measures were visible plaque index, salivary mutans streptococci counts and salivary lactic acid production. The authors observed significant reduction in visible plaque in both aroups after 4-weeks and also diminished lactic acid formation. But the proportion of mutans streptococci decreased significantly only in the test group. The alterations in the test group seemed to be most prominent in children without previous experience.

Badet et al. (2008)[33] had studied the inhibitory effect of xylitol on the formation of an experimental model of oral biofilm at different concentratiQn. Six bacterial strains (Streptococcus mutans ATCC 25175, Streptococcus sobrinus ATCC 33478, Lactobacillus rhamnosus ATCC 7469, Actinomyces viscosus ATCC 15987, Porphyromonas gingivalis ATCC 33277 and Fusobacterium nucleatum ATCC 10953) were used for making biofilms on hydroxyapatite (HA) discs. Xylitol was tested at two concentrations, 1% and 3% and the control biofilms were treatedwith .physiological saline. The authors observed only a small amount of isolated bacteria on the surface of HA discs and inhibition in the growth of different species. By this study, the authors concluded that xylitol has a clear inhibitory effect on the formation of the experimental biofilmsand acid production of cariogenic bacteria and confirms the relevance of using this polyol for the prevention of oral diseases caused by dental plaque.

Blicks et al. (2008)[34] had investigated the effects off lozenges containing either xylitol or xylitol/fluoride on proximal caries development in young adolescents with high caries risk. It was a 2-year double-blind randomized trial with two parallel arms. One hundred and sixty healthy 10-12 year old children with high caries risk were selected. After taking informed consent, the subjects were randomly assigned to two study groups; A) Xylitol, and B) Xylitol/fluoride and were instructed to take two tablets three times a day. Caries risk was assessed by the regular dentists using a software program within the digital records. The authors observed no statistically significant differences in caries incidence. There was somewhat higher prevalence in the xylitol/fluoride group and similar incidence of approximal enamel lesions and total approximal DMFS. By their results the authors concluded by not supporting self-administration of xylitol or xylitol/fluoride containing lozenges for the prevention of approximal caries in young adults with high caries risk.

Biria et al. (2009)[11]compared the effect of mastic chewing gum on remineralization of caries-like   lesions. Cross-over, single blinded, in-situ study approved by the ethics committee of research deputy of ShahidBeheshi Dental School was conducted. Artificial caries like lesions were created in six extracted human premolars, which were cut into 100 microns thick axial sections. Fifteen 20-30years of age healthy volunteers were participated in the study. In the first phase of the two phase study, subjects ctiewed five sticks of gum(mastic gum or xylitol chewing gum) per day, each for 20 minutes, followed by a week of the study construction. They observed that the decrease in demineralized surfaces in both groups was statistically significant, but the difference of average decrease was not statistically significant. The authors concluded that remineralizing potential is due to increased saliva stimulation but no significant difference existed between the effect of mastic gum and xylitol gum.

Fontana et al. (2009)[35] had studied to examine the effects of xylitol gum (XG) on the acquisition pattern of 39 bacterial species including Mutans streptococci (MS), in infants. Ninety-seven mothers (MS counts > 105CFU/m1) who were enrolled for the study were randomly divided into 4 groups and received: 1) XG(4.2gm/day); 2) XG(6months after baseline exams); 3)Sorbitol gum(4.2gm/day); 4)no gum. Groups 1 and 3 chewed gum 3 times a day for 9 months. Micro biota of plaque and saliva samples from the mother and child pairs were analyzed by culturing and via checker-board DNA-DNA hybridization. The authors isolated 33% of MS from pre-dentate infant (<=5months) baseline saliva samples and at final visit they isolated 41% from the saliva and 65% from the plaque samples. At the final visit (9 months later), the authors observed no significant differences between treatment groups for infants 39 microbial plaque species, including MS. By these results, it can be concluded that maternal use of xylitol gum did not result in statistically significant differences in the microbial plaque composition of 9- to 14-month's old infants.

Milgrom et al. (2009)[36] had evaluated the effectiveness of a xylitol pediatric topical oral syrup in reducing the incidence of dental caries among very young children and also in reducing acute obits media in a subsequent study. It was a double-blind randomized control trial. One hundred eight children aged 9-15 months were screened of which 84 completed the study. They compared 2 active treatment groups, each receiving xylitol syrup (8g/d) orally divided into 2 doses or 3 doses vs a control group(2.67g of xylitol). The primary outcome end point of the study was the number of decayed primary teeth. The authors observed that fifteen of 29 of the children in the control group (51.7%) had tooth decay compared with 13 of 32 children in xyl-3X group (40.6%) and eight of 33 children in the xyl-2X group (24.2%). The mean (SD) numbers of decayed teeth were 1.9(2.4) in the control group, 1.0(1.4) in the xyl-3X group, and 0.6(1.1) in the xyl-2X group. Fewer decayed teeth were seen in xyl-2X group and in the xyl-3X group when compared with control group. No statistical difference was noted between the 2 xylitol treatment groups. The authors concluded that 8gm of Xylitol administered typically 2 or 3 tithes daily prevents early childhood caries effectively.

Thaw boon et al.(2009)[37] had determined the remineralization effects of xylitol chewing gum containing funeral and calcium hydrogen phosphate on enamel subsurface lesions in humans. The study was a double-blind, randomized, cross-over design. Four types of gum: (1) Xylitol gum,(2) Xylitol gum containing funoran and calcium hydrogen phosphate, (3) Sugar gum and (4) Gum base as control were used. Subjects were instructed to wear removable lingual appliances, with half-slab insets of human enamel containing demineralized subsurface lesions. They were told to chew at a natural chewing frequency for 20 minutes, 4 times daily for 7 days. It was observed that the mean area of remineralization(AZd-

A.Zr) and mean percent remineralization(%R) in those chewing xylitol gum containing funoran and calcium hydrogen phosphate were significantly higher than the corresponding values for xylitol gum, sugar gum and gum base. It is concluded from the study that xylitol containing funoran and calcium hydrogen phosphate results in significant remineralization of initial caries-like lesions of the teeth.

Hildebrandt et al. (2010)[38] had compared the ability of a xylitol mouth rinse and xylitol chewing gum, used in clinically realistic manners, to reduce mutans streptococci (MS) levels on the teeth of adults. 107 individuals of 202 potential subjects screened for sufficient salivary MS had agreed to participate. Subjects were randomly assigned into three groups: an experimental group (N=36) rinsed with 20ml of an aqueous solution of xylitol twice daily for 60 seconds, a positive control group (N=35) chewed two xylitol gum pellets for at least 5 minutes three times daily and a negative control group(N=34) used neither product. At baseline, mean MS levels were 5.6 (0.1) in positive control, 5.4 (0.1) in experimental, and 5.5 (0.1) in negative control groups has been observed. After 3 months, they were observed as 4.4(0.2), 4.4(0.2) and 4.9(0.2) respectively. MS levels tended to be lower in the experimental and positive control groups which is statistically insignificant.

Paula et al. (2010)[39] had assessed the effect of combining 1% chlorhexidine (CHX) with xylitol chewing gum (XYL) on streptococcus mutans(SM) and biofilm levels in 6-8 year old children. It was a randomized, blind experimental study. Based on the selection criteria 82 children presented with early mixed dentition without active carious lesions were randomly divided into four groups, group 1 (G1)(XYL)(n=20): xylitol chewing gum twice a day after breakfast and lunch; group 2(G2) (n=20): xylitol gum + CHX varnish application at the start of the study and after one and two months; group 3(G3)(n=20): CHX varnish as G2; group 4(G4)(n=22): fluoride gel application at the start of the study and after one and two months. The authors observed greater biofilm reduction in G2 and G3, SM levels reduction in all groups. Largest reduction was observedwith XYL+CHX throughout the study period. The authors concluded that XYL+CHX combination was efficient and superior to single treatments in controlling biofilm and suppressing SM levels.

Nakai et al. (2010)[40] had studied to confirm the effects of maternal chewing of xylitol gum at an earlier period(starting at 6th month of pregnancy and terminating when the child was 9 months of age). After screening, 107 pregnant women with high salivary MS were randomized into two groups: xylitol gum(n=56) and no gum(control;n=51) groups. Outcome measures were the presence of MS in saliva or plaque of the children until age 24 months. The authors observed significantly less MS colonization in xylitol group than control-group children aged 9-24 months. The control group children acquired MS 8.8 months earlier than those in the xylitol group. The authors confirmed the effectiveness of maternal early exposure to xylitol chewing Gum on mother-child transmission of MS.

Fernandes et al. (2011)[41] had studied to compare the effect of intensive treatment with chlorhexidine (CHX) gel, containing saccharin or aspartame, in decreasing the mutans streptococcus (MS) levels in deaf children and to evaluate the reappearance pattern of MS over time after the CHX applications in those subjects. It was a prospective, randomized and double blinded study. Eighteen patients, aged 5-10years were divided into two groups, according to the sweetener used with CHX gel: saccharin (n=9) or aspartame (n=9). Baseline microbial data was collected before CHX treatment. CHX gel was applied 4 times on first day and 3 times on second day consequently. Saliva samples were collected after 7,30,60,90 and 120 days. The authors observed significant decrease of MS levels with CHX gel containing aspartame than with saccharin. Gradual reappearance of MS has been observed after CHX treatment. The authors concluded that a new CHX application may be necessary after 60 days with control as there was a large intra-individual variation in the time of MS recolonization.

Simoeset al. (2011)[42]had studied the effect of 1% Chlorhexidine Varnish (CHX) and 40% Xylitol Solution (XYL) on Streptococcus Mutans(SM) and Plaque Accumulation in 2 to 5 year olds. 68 children were selected with medium levels (1×10^3) to very high levels ($>1 \times 10^5$) of streptococcus mutans (MS) in the saliva.

Subjects were divided into 4 groups of 17 children each: (1) CHX; (2) CHX+XYL; (3) XYL; and (4) 0.05% sodium fluoride (F). An assessment of SM levels and plaque indices was done on all children at baseline, 15 days, and at 1, 3, and 6 months.SM levels were determined by the spatula method. Statistically significant reduction of SM counts in all groups was observed, with CHX and F groups showing the greatest effect. It was concluded that 1% CHX varnish associated with 40% xylitol solution does not provide significant suppression of streptococcus mutans counts and reduction of plaque accumulation at any follow-up time points.

Ramanarayanan et al. (2011)[43] had evaluated the in-vitro antifungal activity of fourcommercially available intense sweeteners against Candida albicans, and assessed their minimal inhibitory concentration against it. Four commercially available intense sweeteners, namelySweetexCD (saccharin), Sugarfree gold® (aspartame), Sugarfreenatura® (sucralose), and So sweet® (stevioside), were obtained and powdered. Necessary concentrations ofthe sweeteners were prepared by mixing them with an inert solvent. The authors observed that at the end of 48hrs, statistically significant antifungal activity was demonstrated by all the sweeteners used in this study (P = 0.001). and zone of inhibition demonstrated against C.albicans was maximum. It was observed that So sweet® demonstrated a 'Minimal inhibitory concentration (MIC) of 6.25%. Sugarfree gold® and sugarfreenatura® demonstrated an MIC ofl 2.5% and Sweetex® demonstrated an

MIC of 50%.In conclusion, all the sweeteners used in this study have demonstrated significant antifungal activity. With the proved antibacterial and anti/non-cariogenic properties and safety, these sweeteners could be recommended as an ideal alternative to sucrose.

Schernhammeret al. (2012)[44]conducted a study to evaluate whether the consumption of aspartame and sugar-containing soda is associated with risk of hematopoietic cancers. The authors repeatedly assessed diet in the Nurses' Health Study (NHS) and Health Professionals Follow-Up Study (HPFS). Over 22 y, they identified 1324 non-Hodgkin lymphomas (NHLs), 285 multiple myelomas, and 339 leukemias. Incidence RRs and 95% CIs were calculated by using Cox proportional hazards models. The authors concluded that although the findings preserve the possibility of a detrimental effect of a constituent of diet soda, such as aspartame, on select cancers, the inconsistent sex effects and occurrence of an apparent cancer risk in individuals who consume regular soda do not permit the ruling out of chance as an explanation.

Azzaet al. (2012)[45]conducted a study to evaluate the effect of aspartame intake on the histological and genetic structures of mother albino rats and their offspring. Sixty adult female albino rats and 180 of their offspring were equally divided into two groups (control and treated),each group divided into three subgroups. Each subgroup consisted of 10 pregnant rats and 30 of their offspring. Each pregnant rat in the treated subgroups was given a single daily dose of 1 mL aspartame solution (50.4 mg) by gastric gavage throughout the time intervals of experimental design. At the end of each experimental period for control and treated subgroups, the liver of half of both control and treated groups were subjected for histological study while the liver and bone marrow of the other halves were subjected for cytogenetic studies. The results revealed that the rats and their offspring in the subgroups of control animals showed increases in body weight, normal histological sections, low chromosomal aberration and low DNA fragmentation. The treated animals in the three subgroups rats and their offspring revealed decreases in body weight, high histological lesions, increases in the chromosomal aberration and DNA fragmentation compared with control groups. In conclusion, the consumption of aspartame leads to histopathological lesions in the liver and alterations of the genetic system in the liver and bone marrow of mother albino rats and their offspring. These toxicological changes were directly proportional to the duration of its administration and improved after its withdrawal.

Cong et al. (2013)[46]investigated the peripheral and central nervous system effects of protracted exposure to .a widely used artificial sweetener, acesulfame K (ACK). They found that extended ACK exposure (40 weeks) in normal C57BL/6J mice demonstrated a moderate and limited influence on metabolic homeostasis, including altering fasting insulin and leptin levels, pancreatic islet size and lipid levels, without affecting insulin sensitivity and bodyweight. It was concluded that chronic use of ACK could affect cognitive functions, potentially via altering neuro-metabolic functions in male C57BL/6J mice.

Kashanian et al. (2013)[47]studied whole Aspartame and DNA interaction via different spectroscopic methods. A number of small molecules bind directly and selectively to DNA, by inhibiting replication, transcription or topoisomerase activity. In this work the interaction of native calf thymus DNA (CT-DNA) with Aspartame (APM), an artificial sweeteners was studied at physiological pH. DNA binding study of APM is useful to understand APM—DNA interaction mechanism and to provide guidance for the application and design of new and safer artificial sweeteners. The interaction was investigated using spectrophotometric, spectro-fluorometriccompetition experiment and circular dichroism (CD). Hypochromism and red shift are shown in UV absorption band of APM. A strong fluorescence quenching reaction of DNA to APM was observed and the binding constants (KO of DNA with APM and corresponding number of binding sites (n) were calculated at different temperatures. The authors concluded that APM interacts with calf thymus DNA via groove binding mode with an intrinsic binding constant of 5x 10+4M-1.

Findikli Z, Tiirkoglu S (2014)[48]conducted a study to investigate the genotoxicity of the artificial sweeteners acesulfame potassium, aspartame, saccharin, and sorbitol, which are used in food industry and by patients with diabetes, in human peripheral lymphocyte cells using the single-cell gel electrophoresis (comet) technique. Human lymphocyte cells were treated with the substances for three hour at each of the three dosages (1.25, 2.5, and 5 ppm). The. chemical additives were studied, and the related DNA damages in the study group were compared to the control group for each of the treatment dosages. The DNA breakages observed in the comet assay were assessed in terms of tail moment and tail DNA percent using the comet parameters Based on the results for the short-term in vitro treatments, the 4 different food flavorings were found to have genotoxic effects.

SUGARS AND DENTAL CARIES

SUGARS:

For early man, the ability to differentiate between sweet and bitter substances may have had survival value, since nutritious plants are usually sweet whereas poisonous plants are often bitter. But the modern man who is now living in a habitat which is influenced by multiple factors, is now considering sugar as 'sweet poison' which is directly or indirectly responsible for many of his sufferings.

The etymology of the word "sugar" has been traced to the Chinese term shache, literally, "sand-sugar plant", then to the Sanskrit, sharkera, meaning sand or gravel, and more directly from the Arabic, sukkar.49 Sucrose (table sugar) which is the common house hold food stuff, is referred to as sugar by most of the people. Sucrose is only one of the many naturally occurring sugars used in the human diet. Actually the term 'sugar' is applied to carbohydrates soluble in water and sweet to taste. Thus, monosaccharides and oligosaccharides are considered as sugars.

Global oral epidemiology bank revealed a significant relationship between sugar supplies and dental caries for 12-year old children in 47 populations. In India sugar consumption increased from 9.7kg/person per annum in 1983 to 12.1kg/person per annum in 1999-2000 and sugar demand is expected to increase almost nine-fold by 2026 from base year demand of 11.9 mt (metric tonnes) in 1999-2000.Sugar consumption has been declining consistently in both rural and urban areas, but continues to be still higher in urban India.

SUGARS AND DENTAL CARIES:

Dental cariesstill remains a very costly and widespread diseasethat in many industrialized countries it affects mainlydisadvantaged individuals and is of serious concern inmany developing countries. One of the earlier references to the connection between sugar and dental caries was in 1598, when Hentzer, a German traveller ascribed blackness of Queen Elizabeth's teeth was due to excessive consumption of sugar and sugar dishes. During world-war II, when there was a reduction of sugar production, caries incidence was less than the pre-and post-war periods.Sugar (sucrose) being most acceptable sweetening agent in use by mankind is considered as the 'Arch Criminal' in dental caries initiation.56 Man consumes sugar in some form or the other from infant age before eruption of teeth into the oral cavity. Tooth decay is the most prevalent childhood chronic disease and is 5 times more common than asthma. Children with early childhood caries (ECC) are 3 times more likely to develop caries in permanent teeth than children without ECC.

SUGAR AS AN ETIOLOGICAL FACTOR IN DENTAL CARIES

Many theories have evolved through years in explaining the etiology of dental caries, which is an interplay between oral bacteria, local carbohydrates and tooth surface that may be shown schematically as follows:

Bacteria + Sugars + Teeth -÷ Organic acids Dental caries

But according to current concepts, dental caries is a multi-factorial disease where 'time' factor is also considered . However, it is also known that these four criteria are not always enough to cause the disease and a sheltered environment promoting development of a cariogenic biofilm is required. The caries process does not have an inevitable outcome, and different individuals will be susceptible to different degrees depending on the shape of their teeth, oral hygiene habits, and the buffering capacity of their saliva. Dental caries can occur on any surface of a tooth that is exposed to the oral cavity, but not the structures that are retained within the bone.

The bacteria most responsible for dental cavities are the mutans streptococci, most prominently Streptococcus e mutans and Streptococcuss obrinus, and lactobacilli. If left untreated, the diseasecan lead to pain, tooth loss and infection. Tooth decay disease is caused by specific types of bacteria that produce acid in the presence of fermentable carbohydrates such as sucrose, fructose, and glucose. The mineral content of teeth is sensitive to increases in acidity from the production of lactic acid. To be specific, a tooth (which is primarily mineral in content) is in a constant state of back-and-forth demineralization and remineralization between the tooth and surrounding saliva. These patients may be susceptible to dental caries. When the pH at the surface of the tooth drops below 5.5, demineralization proceeds faster than remineralization (meaning that there is a net loss of mineral structure on the tooth's surface).

All caries occur from bacterial acid demineralization that exceeds saliva and fluoride remineralization, and acid demineralization occurs where bacterial plaque is left on teeth. Some foods have an acidic pH of 5.5 or lower which can result in demineralization in the absence of bacteria. This is known as erosion, rather than caries, because the acid is not bacterial in origin.

Bacteria in a person's mouth convert glucose, fructose, and most commonly sucrose (table sugar) into acids such as lactic acid through a glycolytic process called fermentation. If left in contact with the tooth, these acids may cause demineralization, which is the dissolution of its mineral content. The process is dynamic, however, as remineralization can also occur if the acid is neutralized by saliva or mouthwash. Fluoride toothpaste or dental varnish may aid remineralization. If demineralization continues over time, enough mineral content may be lost so that the soft organic material left behind disintegrates, forming a cavity or hole. The impact such sugars have on the progress of dental caries is called cariogenicity. Sucrose, although a bound glucose and fructose unit, is in fact more cariogenic than a mixture of equal parts of glucose and fructose. This is due to the bacteria utilizing the energy in the saccharide bond between the glucose and fructose subunits. S.mutans adheres to the biofilm on the tooth by converting sucrose into an extremely adhesive substance called dextran polysaccharide by the enzyme dextran-sucranase.

Exposure: The frequency of which teeth are exposed to cariogenic (acidic) environments affects the likelihood of caries development. After meals or snacks, the bacteria in the mouth metabolize sugar, resulting in an acidic by-product that decreases pH. As time progresses, the pH returns to normal due to the buffering capacity of saliva and the dissolved mineral content of tooth surfaces. During every exposure to the acidic environment, portions of the inorganic mineral content at the surface of teeth dissolves and can remain dissolved for two hours. Since teeth are vulnerable during these acidic periods, the development of dental caries relies heavily on the frequency of acid exposure.

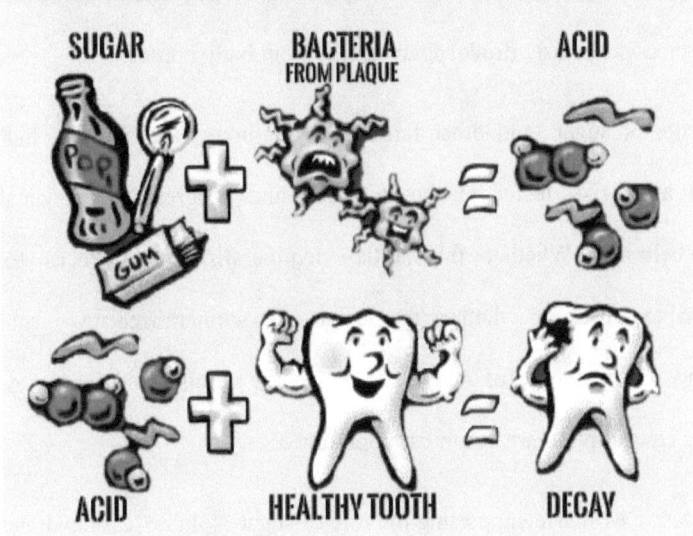

HOW SUGAR CAUSES TOOTH DECAY?

SUGAR   BACTERIA FROM PLAQUE   ACID

ACID   HEALTHY TOOTH   DECAY

The carious process can begin within days of a tooth's erupting into the mouth if the diet is sufficiently rich in suitable carbohydrates. Evidence suggests that the introduction of fluoride treatments have slowed the process. Proximal caries take an average of four years to pass through enamel in permanent teeth. Because the cementum enveloping the root surface is not nearly as durable as the enamel encasing the crown, root caries tends to progress much more rapidly than decay on other surfaces. The progression and loss of mineralization on the root surface is 2.5 times faster than caries in enamel. In very severe cases where oral hygiene is very poor and where the diet is very rich in fermentable carbohydrates, caries may cause cavities within months of tooth eruption. This can occur, for example, when children continuously drink sugary drinks from baby bottles.

The role of sugar (and other fermentable carbohydrates such as highly refined flour) as a risk factor in the initiation and progression of dental caries is overwhelming. Whether this initial demineralization proceeds to clinically detectable caries or whether the lesion is re-mineralized by plaque minerals depends on a number of factors, of which the amount and frequency of further sugars consumption are of utmost importance.

The classic evidence supporting the role of sugar (soluble carbohydrates) in dental caries in man is well documented by some of the studies like— The Vipeholm Study, Turku Sugar Study, World War II Food Rationing, Hopewood House Study, Tristan da Cunha, Hereditary Fructose Intolerance, Experimental Caries in Man, and Stephan Plaque pH Response and are listed in Table-1.59 Table 2 provides a partial list of review articles that have looked at the role of sugars in dental caries, with most authors supporting the relationship.

Table 1: Classic evidence from humans supporting the role of sugar in dental caries

STUDY	REFERENCE(s)	MAIN CONCLUSIONS
Vipehlom study	Gustafsson et al [1954]	The more frequently sugar is consumed the greater the risk;sugar consumed between meals has much greater caries potential than when consumed during a meal
Turku sugar	Scheinin et al [19756]	When sugars are almost completely replaced by non-fermentable sugar substitutes,caries increment is dramatically reduced;fructose is less cariogenic than sucrose
World WarII	Toverud [1957 a,b] Takeuchi[1961]	Caries decreased and increased with sugar consumption during and after the war respectively
Hopewood house	Harris [1963]	Modern diet more cariogenic than vegetarian low sugar diet
Tristan da Cunha	Hollowat et al[1963] Fisher[1968]	Introduction of a modern diet including sugar and refined carbohydrate to this remote island greatly increased caries prevalence
Hereditary Fructose Intolerance	Marthaler [1967] Newbrun et al[1980]	Less caries in individuals that must avoid sucrose and fructose,but not other sugars and complex carbohydrate
Experimental Caries in Man	von der Fehr et al[1970] Geddes at al [1978]	Incipient caries can be rapidly induced by frequent rinsing with high concentration sucrose solutions in the absence of oral hygiene
Stephan plaque pH response	Stephan [1940,1944]	Demonstrated the relationship between sugar exposure resulting in the acidification of dental plaque and caries experience

Table 2: Review articles on the relationship between sugar (diet) & dental caries

AUTHOR(s)	MAIN CONCLUSION
Marthaler[1967]	Foodstuff containing simple sugars are far more cariogenic than common starchy food
Newbrun [1969]	Called for the specific elimination of sucrose or sucrose containing foods rather than restricting total carbohydrate consumption
Bibby [1975]	Snack foods share importance with sucrose in caries causation
Sreebny [1982a]	Total consumption and frequency of intake contribute to dental caries;lacking evidence about the precise definition of the relationship
Newburn [1982a]	Compelling evidence that the proportion of sucrose in a food is one important determinant of its cariogenicity
Sheiham [1983]	Sugar is the principal cause of caries in industrialized counteies;recommended that sugar consumption be reduced to 15kg/person/year or below
Shaw [1983]	Studies in animals consistent with the clinical evidence on the relationship between sugar and caries
Rugg-Gunn[1986]	Cariogenicity of staple starchy food is low;the addition of sucrose to cooked starch is comparable to similar quantities of sucrose;fresh fruits appear to have low cariogenicty
Bowen & Birkhed[1986]	Frequency of eating sugars is of greaer importance than total sugar consumption
Walker & Cleaton-Jones[1989]	Degree of incrimination of sugar as a cause of caries is grossly exaggerated;questioned predictions of reductions in caries from decreases in sugar and snack intakes
Marthaler [1990]	In spite of dramatic reductions in caries due primarily to widespread use of fluoride,sugar continue to be the main threat to dental health

Rugg-Gunn[1990]	Dietary modification involving restriction on the frequency and amount of extrinsic sugars can be more effective than othercontrol measures
Konig & Navia [1995]	Acknowledged the relationship between frequency and sugar intake and caries;recommended removing the focus away from elimination of sugar and towards improved oral hygiene and use of fluoride toothpaste
Ruxton et al[1999]	Evidence strongly supports formulation of advice on frequency of consumption,not amount
Konig [2000]	Dental health problems do not require any dietary recommendations other than those required for maintenance of general health
van Loveren [2000]	If good oral hygiene is maintained and fluoride is supplied frequently,teeth will remain intact even if carbohydrate containing food is frequently eaten
Sheiham [2001]	Sugars,particularly sucrose,are the most important dietary cause of caries;the intake of extrinsic sugars greater than 4times a day increases caries risk;sugar consumption should not exceed 60g/day for teenagers and adults and proportionally less for younger children

SUGAR SUBSTITUTES & CLASSIFICATION

As a result of urbanization, man is accustomed to sedentary life which is the cause for many diseases like diabetes, obesity and cardiovascular diseases. In excavating the causes for these conditions, he came across an alternative for sugars i.e., sugar substitutes which provide zero or low calories. But now-a-days due to increased per-capita income and increased awareness about health and physique consciousness and influencing advertisements, these sugar substitutes have become a part of daily menu.

The search for a suitable sugar substitute, or non-nutritive sweetening agent, was Originally prompted by the requirement of diabetic patients for a tasty diet. The dental profession shares an interest in the search for safe, palatable sugar substitutes. The problem of finding a sugar substitute is not simply that finding a sweetener, but an ideal sugar substitute must have some of the physical properties found in sugar.

CLASSIFICATION OF SUGAR SUBSTITUTES

A sugar substituteis a food additive that duplicates the effect of sugar in taste, usually with less food energy. Some sugar substitutes are natural and some are synthetic.: ost of the sugar substitutes are the derivatives of water soluble, crystalline carbohydrates (monosaccharide's and oligosaccharides). Sugar substitutes can be classified in mainly in two ways.

I.   Based on the energy supply into carbohydrate (caloric) and non-carbohydrate(non-caloric)

II. Based on the availability into natural and synthetic

The terms synthetic and artificial sweeteners are no longer used by food technologists because the important difference between sugars and related polyols and the non-caloric sweeteners is a nutritional one. Comparison of sweetness based on energy content is also not meaningful as many of these have little or no food energy.

The sweetness and energy of sugar substitutes in composition to those of sucrose are given below:

NATURAL SUGAR SUBSTITUTES:-

NAME	SWEETNESS BY WEIGHT	SWEETNESS BY FOOD ENERGY	ENERGY DENSITY	NOTES
Brazzein	800			Protein
Curculin	550			Protein
Erythritol	0.7	14	0.05	
Glycyrrhizin	50			
Glycerol	0.6	0.55	1.075	E422
Hydrogenated	0.4-0.9	0.5-1.1	0.75	

starch				
Isomalt	0.45-0.65	0.9-1.3	0.5	E953
Lactitol	0.4	0.8	0.5	E966
Luohanguo	300			
Mabinlin	100			
Maltitol	0.9	1.7	0.525	E965
Monellin	3000			Protein
Osladin				
Pentadin	500			Protein
Sorbitol	0.6	0.9	0.65	E420
Stevia	250			
Tagatose	0.92	2.4	0.38	
Thaumatin	2000			E957
Xylitol	1.0	1.7	0.6	E967

ARTIFICIAL SUGAR SUBSTITUTES:-

NAME	SWEETNESS BY WEIGHT	TRADE NAME	FDA approval	NOTES
Acesulfame potassium	200	Nutrinova	1988	E950
Alitame	2000			
Aspartame	160-200	Nutra Sweet	1981	E951
Salt of aspartame	350	Twin sweet		E962
Cyclamate	30		Banned 1969	E952
Dulcin	250			Banned 1950
Glucin	300			
Neotame	8000	Nutra Sweet	2002	

P-4000	4000		Banned 1950	
Saccharin	300	Sweet N Low	1958	E954
Sucralose	600	Kaltame	1998	E955

SUGAR SUBSTITUTES IN DENTISTRY

Dentists are placed in a difficult situation, as diet control for caries control is not a practicable means of caries prevention as they must compete with the food manufacturers' marketing techniques such as advertisements. The major factors that prompted for the search of suitable sugar substitutes related to dental health are as follows.

1. The attempt to persuade the patients to adopt special dietary programs to limit the frequency with which sugar-containing food are ingested could not be practically achieved for the prevention of caries on a public health scale. It is indeed difficult to change dietary habits, especially if the change requires the elimination of palatable, conveniently available food.

2. Animal experimentation clarified many aspects of cariogenicity of sucrose relative to other dietary components and led to the discovery that in most instances dental decay is a sucrose-dependent infection involving Streptococcus mutans. The observations in the human have associated S.mutans with human dental decay and generally support the sucrose/S. mutans interactions observed in the animal models. As the emergence of S. mutans appears to be sucrose dependent, then tactics which reduce sucrose bioavailability in the plaque ecosystem should constitute effective preventive and/or therapeutic measures in caries control.

Interferences with between meal sucrose bioavailability can be most easily achieved by the use of sugar substitutes, and this provides the rationale for prevention through their use.

3. The Turku xylitol chewing gum study. The clinical trials conducted in Turku showed that xylitol was non-cariogenic and quite possibly anti-cariogenic when substituted for sucrose either in foods or in chewing gums. The xylitol food study was a 2-year clinical trial in which young adults volunteers consumed food sweetened with sucrose, or fructose, or xylitol. The xylitol group exhibited at the end of the study about an 85-90% reduction in caries scorecompared to the sucrose group.

4. Cariogenicity of S. mutans is based upon it having a pH optimum at the critical pH for enamel demineralization. So, one can simultaneously discriminate against S. mutans and promote enamel demineralization by procedures which prevent or minimize acid production in the plaque. In this regard the use of sucrose substitutes becomes an extremely attractive tactic for the control and/or prevention of dental caries. The potential of this approach was first realized by the remarkable 80% reduction of caries relative to a sucrose control observed with a xylitol chewing gum in the Turku study.

HIGH-INTENSITY SWEETENERS

These sweeteners are used to replace sugars in tabletop sugar, which are generally sweeter than sucrose. They may also becalled alternative, artificial, high-intensity, or non-nutritive sweeteners, can replace the sweetness of sugar while providing few or no calories. In addition to the calorie savings, these sugar substitutes have the advantage of not promoting tooth decay, and they are useful in dietary planning for people who are coping with obesity or diabetes. In using table sugar, intensive sweeteners are added to a bulk non-digestible polysaccharide. They are used in the food and drink industry and in medicinal preparations, dentifrices, and mouthwashes. These can be divided into chemically synthesised sweeteners, including saccharin, aspartame and sucralose, and those obtainedfrom plants, including stevioside, thaumatins, and monellin.

Lead acetate (sometimes called sugar of lead), which is an artificialsugar substitute made from lead that is of historical interest because ofits widespread use in the past, such as by ancient Romans. The use oflead acetate as a sweetener eventually produced lead poisoning in anyindividual ingesting it habitually. Lead acetate was abandoned as afood additive throughout most of the world after the high toxicity oflead compounds became apparent.

Acesulfame-K:

In 1967, in the laboratories of Hoechst AG, it was found by chance that compounds with the dihydro-oxathiazinone dioxide ring-system had a sweet taste. It is inconspicuous because it is almost always used in combination with other sweetening agents. When used in this way, it contributes to creating a sweet taste very close to that of sugar. However, if used alone, it can have a bitter aftertaste that consumers would find undesirable. Acesulfame-K is approximately 200 times as sweet as sugar, and it provides zero calories.

More than 50 studies of various aspects of safety were conducted before the FDA approved Acesulfame-K (fig 3) for use in dry foods in 1988, and additional tests were conducted before FDA approved its use in beverages a few years later.Acesulfame-K, sold under the brand name Sunett, is the most successful sugar substitute that you've probably never heard of. Safety studies have found no evidence of carcinogenicity, mutagenicity, cytotoxicity, or teratogenicity.

Recent, re-evaluations of the scientific evidence on acesulfame-K, including acomprehensive review by the food safety authorities of the European Union in 2000, have reaffirmed its safety. No human health problems associated with the consumption of acesulfame-K have been reported in the scientific literature, despite more than 15 years of extensive use in many countries.

Aspartame:

Aspartame was discovered in 1965 and approved by the FDA in 1981. Aspartame is a dipeptide ester in which aspartic acid is bound at the N-terminal of phenylalanine. It is an odourless white crystalline powder with a refreshing sweet taste, and in a 4% aqueous solution and is about 180 times sweeter than sugar. During the first years after approval, when aspartame was sold exclusively by the patent holder, it was known primarily by the brand names NutraSweet and Equal (the latter is the popular table-top sugar substitute sold in blue packets). Aspartame itself does not occur naturally.

Although the quality of taste of aspartame is not as 'mellow' as that of sucrose, it resembles that of sorbitol and is less bitter and stringent than that of stevioside. Aspartame is stable in a powder state at lower temperatures, but polymerisation occurs at temperatures exceeding 100°C. It is most stable at pH 4, and at this pH even heating at 100°C for 60 minutes in an aqueous solution results in minimal decomposition, compared with the same heating conditions at pH 6 when 90% or more will be decomposed. The available energy value is 4kcal/g. However, since only a small amount is routinely used in food, the calorific value is negligible.

Unlike most other low-calorie sugar substitutes, aspartame is broken down inthe human body. Enzymes in the digestive tract break it down into its components (phenylalanine, aspartic acid, and methanol), each of which is then metabolized just as it would be if derived from other dietary sources. Because aspartame is metabolized, it provides as many calories as an equivalent weight of protein or carbohydrate does. However, because aspartame is intensely sweet, the amount used in foods and beverages is so small that its caloric contribution is negligible.

Foods and beverages that contain aspartame must carry a label statement indicating that the product contains phenylalanine. This statement is for the benefit of individuals with the disease phenyl ketonuria, who must strictly limit their intake of this amino acid. Phenylketonuria is a rare disease, affecting approximately one in 15,000 people, that results from a hereditary lack of an enzyme necessary for the normal metabolism of phenylalanine. Unless the disorder is detected in early infancy and treated with a phenylalanine-restricted diet, it results in mental retardation and other severe, permanent effects. Newborn infants in the U.S. and many other countries are screened for phenyl ketonuria at birth. Because of screening and effective treatment, substantial numbers of people with phenyl ketonuria are living near-normal lives except for the need for dietary restriction.

Aspartame is unstable if subjected to prolonged heating and therefore cannot be used in baking or cooking (unless added at the end of the cooking process). Aspartame also decomposes in liquids during prolonged storage (this is why diet soft drinks have a shelf life about half that of regular soft drinks). The relative instability of aspartame is a quality issue, not a safety issue. For example, if you drink a can of diet soft drink that has been left too long in a hot car, causing some of the aspartame in the beverage to break down, it will notmake you sick. However, you may notice deterioration in the quality of the beverage.

Neotame:

Neotame is the newest of the low-calorie sugar substitutes. It was approved in 2002 and has not yet appeared in commercial products in the United States. Like aspartame, neotame contains the amino acids phenylalanine and aspartic acid. The two amino acids, however, are combined in a way that is different from that in aspartame, giving neotame different properties. Neotame is extraordinarily sweet, with sweetness potency at least 7,000 times that of sugar and at least 30 times that of aspartame. Unlike aspartame, neotame is heat stable and therefore can be used in cooking and baking.

Neotame
(neohexyl-aspartame)

Neotame is chemically similar to aspartame. It had to be comprehensively tested for safety, just as any other new food additive would, before it was approved by the

FDA. Thescientific evidence submitted to FDA by neotame's manufacturer in support of its safety included the results of more than 110 scientific studies, including tests in both experimental animals and human volunteers. Safe for use by the general population, including pregnantana lactating wuiucu, children, and people with diabetes.

Neotame is broken down into a derivative and methanol, both of which are rapidly excreted from the body through either the digestive tract or the urinary tract. Although neotame containsphenylalanine, products sweetened with neotame will not be required to bear a warning notice for people with phenylketonuria, as the amount of phenylalanine in a neotame-sweetened product is so small that it is insignificant, even for people who must limit their phenylalanine intake.

Saccharin:

Saccharin, the oldest low-calorie sugar substitute, was discovered in 1878. It is 300 times sweeter than sugar and provides no calories. In the first half of the twentieth century, saccharin was popular as a sugar substitute in the diets of people with diabetes and other medical conditions. It was also used extensively as a replacement for strictly rationed sugar in Europe during both World Wars. Between 1970 and 1981, saccharin was the only low calorie sugar substitute available in the United States. Saccharin is still widely used today, often in combination with other sugar substitutes, and owes much of its popularity to its low cost. Although saccharin can have a bitter aftertaste when used alone, it works well in blends with other sugar substitutes.

During the 1970s, concerns were raised about whether saccharin might be capable of causing human cancer. In several studies in which a particular chemical form of saccharin, sodium saccharin, was administered to rats in extremely large doses for a lifetime, the male rats had an increased rate of bladder cancer. In 1977, on the basis of this evidence, the FDA attempted to ban saccharin, which met with an extremely negative reaction from the American public as saccharin was the only low-calorie sugar substitute on the market at that time. Acting in response to a massive public mandate, Congress passed a law that imposed a moratorium on the proposed FDA action, and saccharin was never banned, although a warning label was required on saccharin-sweetened products.

Since the 1970s, scientific research has shown that saccharin is not a cancer hazard in humans as this phenomenon does not occur in humans, whose bladder physiology is quite different from that of rats. The relationship between saccharin and bladder cancer has been evaluated in epidemiological studies (studies of the occurrence of disease in human populations), most of which used the case-control design (i.e., people diagnosed with bladder cancer were compared with people of the same age and sex who did not have the disease to see how their past experiences, including exposure to saccharin, differed), in which no detectable association between saccharin consumption and the risk of bladder cancer in humans was observed. Regulatory agencies and international organizations have removed saccharin from their lists of probable human carcinogens, as there is no relevant evidence of cancer hazard from the use of saccharin and the requirement for a warning label on saccharin-sweetened products has been discontinued. There are no unresolved safety issues pertaining to saccharin at the present time.

Sucralose:

Sucralose was discovered in 1976. It is made from sucrose (table sugar) by a process that substitutes three chlorine atoms for three hydrogen-oxygen (hydroxyl) groups on the sucrose molecule. It has no nutritional value and is non-caloric. Although sucralose is made from sugar,the human body does not recognize it as a sugar and does not obtain energy by breaking it down;in fact almost all of it excreted from the body unchanged.Sucralose is about 600times sweeter than sugar,and it is heat-stable.Like the other low calore sugar substitutes,it does not promote tooth decay. In U.S. it is sold under the brand name Splenda and is perhaps most familiar to U.S. consumers as the sugar substitutes that comes in yellow packets.

Stevia sweetener: (Stevioside, Rebaudioside)

This sweetener is extracted from the leaf of stevia (Compositae) (Stevia rebaudianaBertoni) , which is harvested in the highlands'of Paraguay and other parts of South America, and its main components are steviosides. Three types of stevia sweeteners exist. The regularproduct , consisting mainly of a stevioside, the Reva A, consisting mainly of rebaudioside A,and the sugar metastasis product. In the regular product, the content ratio of stevioside to rebaudioside ranges from 7:3 to 8:2, while in the Reva A this ratio is about 1:3. Since rebaudioside has a very sweet taste, the quality of sweetness of Reva A is higher than that of the regular product.

The degree of sweetness of stevia is between 150 and 300 times that of sucrose. The majority of the ingested stevia sweetener is utilised by enterobacteria, and the remainder is excreted in the stools. The available energy value is 0 kcal/g.

Results from several studies have shown stevia sweeteners to be non-cariogenic. In animal caries experiments, significant differences were found in the sulcal caries scores and S. sobrinus counts between the sucrose group and the stevia sweeteners group. There were no significant differences between the stevioside and rebaudioside A. This study concluded that neither stevioside nor rebaudioside A is cariogenic.

Structural formula of Stevioside

Thaumatin:

Thaumatin is a mixture of intensely sweet proteins (thaumatins) extracted with water from the arils of the fruit of the West African perennial plant Thaumatococcusdaniellii. The thaumatins have a normal complement of amino acids, except that histidine is not present. The molecular weights of the thaumatins are approximately 22,000. It is 2,000 times as sweet as sucrose. Although it has been confirmed that mutans streptococci does not liberate acid or insoluble glucan from thaumatin, there have been no other studies on anti-cariogenic nature.

Monellin:

Monellin is a sweet protein extracted from African serendipity berries, Dioscoreophylhuncumminsii. This protein has two polypeptide chains of 45 and 50 amino acids and is about 70,000 times sweeter than sucrose. The sweet sensation persists in the mouth for an unusual length of time. Although monellin had no growth activity of cariogenic bacteria, there have been no other studies on anti-cariogenic nature.

Cyclamate:

Cyclamate was discovered accidentally by Michael Sveda in 1937 and was used as a sugar substitute in about 50 countries. Cyclamate is not a new product; it in the U.S. in the1950s and 1960s, primarily in a very successful blend with saccharin. In

1970, however, cyclamate was banned in the U.S. in response to an animal experiment that seemed to indicate that it could cause bladder cancer. Later, extensivefurther studies in several animal species did not show any link between cyclamateand cancer. Thus, on the basis of the complete body of evidence, scientistshave concluded that cyclamate is not a cancer-causing agent. The manufacturerof cyclamate has submitted a petition for its reapproval in the United States.This petition, like the one for alitame, is currently being "held in abeyance" (as of March 2006) while additional scientific data are developed.

Dihydrochalcone sweeteners:

Neohesperidinedihydrochalcone (NHDC) is a glycosidicchalcone which is an. artificial intense sweetener, 1000-1800 times as sweet as sucrose. NHDC has not been reported to occur naturally in any food items but neohesperidine, the raw material for neohesperidinedihydrochalcone, is extracted from the immature fruit of Citrus auruntizun(bitter orange), or obtained from naringin, the main flavonoid in Citrus paradisii(grapefruit).

The absorption of NHDC is associated with deglycosylation leading to the aglyconehesperetindihydrochalcone, followed by the formation of glucuronate and sulphate conjugates, which are excreted as such with the urine and bile. Since the absorption of flavonoids is typically incomplete, there is a substantial microbial degradation of unabsorbed flavonoids in the human intestinal tract yielding, in the case of NHDC, phloroglucinol and ferulate.

Miraculin:

Miraculinis a glycoprotein extracted from the miracle berry ormiracle fruit plant, a shrub native to West Africa (Synsepahunclulcificumor Richacielladulcifica). Miraculin itself is not sweet, but has taste-modifying function of the fruit had been regarded as a miracle. The active substance, isolated by Prof KenzoKurihara (a Japanesescientist in 1968). Miraculinoccurs as a tetramer , a combination of 4 monomersgroup by dimer. Within each dimer 2 miraculin glycoproteins arelinked by a disulfide bridge.

The detailed mechanism of the taste-inducing behaviour is still unknown,but has been suggested that the miraculin protein can changethe structure of taste cells on the tongue. As miraculin is a readily soluble protein and relatively heatstable, it is a potential sweetener in acidic food (e.g. soft drinks). Miraculin was never approved for use as a sweetener by the United States Food and Drug Administration (FDA) and has no legal status in the European Union. How ever it is approved in Japan as a harmless additive.

Glycyrrhizin:

Glycyrrhizin is the main sweet-tasting compound from liquorice root. It is 30 — 50 times as sweet as sucrose (table .sugar).Pure glycyrrhizin is odorless. Chemically glycyrrhizin is a triterpenoids aponinglycoside of glycyrrhizic (or glycyrrhizinic) acid. Upon hydrolysis, the glycoside loses its sweet taste and is converted to the aglyconeglycyrrhetinic acid plus two molecules of glucuronic acid. In the United States, glycyrrhizin is classified as "generally recognized as safe"(GRAS) as a flavoring agent, although not as a sweetener. Glycyrrhizin is used as flavouring in some candies, pharmaceuticals, and tobacco products.

Dulcin:

Dulcin is also known by the names sucroland valzin. It is an artificial sweetener about 250 times sweeter thansugar discovered in 1884 by Joseph Berlinerbau . It was first massproducedabout seven years later. Despite the fact that it wasdiscovered only five years after saccharin, it never enjoyed the lattercompound's market success. Still, it was an important sweetener ofthe early 20th century and had an advantage over saccharin in that itdid not possess a bitter aftertaste. Early medical tests marked the substance as safe for humanconsumption, and it was considered ideal for diabetics. However, anFDA study in 1951 raised many questions about its safety resulting inits removal from the market in 1954 after animal testing revealed unspecified carcinogenic properties.

High-fructose corn syrup (HFCS):

HFCS is alsocalled iso glucose, maize syrup, or glucose-fructose syrup in the UK, and glucose /fructose in Canada — comprises any of a group of corn syrups that has undergone enzymatic processing to convert its glucose into fructose and has then been mixed with pure corn syrup (100 % glucose) to produce a desired sweetness. In the United States, HFCS is typically used as a sugar substitute and is ubiquitous in processed foods and beverages, including soft drinks, yogurt, industrial bread, cookies, salad dressing, and tomato soup. The most widely used varieties of high-fructose corn syrup are: HFCS 55 (mostly used in soft drinks), approximately 55% fructose and 45% glucose; and HFCS 42 (used in many foods and baked goods), approximately 42 % fructose and 58 % glucose. HFCS — 90,approximately 90 % fructose and 10 % glucose, is used in small quantities for specialty applications, but primarily is used to blendwith HFCS 42 to make HFCS 55.

Since its introduction. HFCS has begun to replace sugar invarious processed foods in the United States, as it is cheaper and is easier to blend and transport because it is a liquid.Several studies funded by Tate & Lyle, a large corn refiner, the American Beverage Institute and the Corn Refiners Association, have defended HFCS.

SUGAR ALCOHOLS (polyols)

A sugar alcohol (also known as a **polyols, polyhydric alcohol, or poly alcohol**) is a hydrogenated foam of carbohydrate, whose carbonyl group (aldehyde or ketone, reducing sugar) has been reduced to a primary or secondary hydroxyl group (hence the alcohol). Sugar alcohols have the general formula H(HCH0)$_{\ldots}$111, whereas sugars have H(HCH0)$_{\ldots}$HCO. In commercial foodstuffs sugar alcohols are commonly used in place of table sugar (sucrose), often in combination with high intensity artificial sweeteners to counter the low sweetness. Of these, xylitol is perhaps the most popular due to its similarity to sucrose in visual appearance and sweetness. Sugar alcohols do notcontribute to tooth decay. Sugar substitutes discussed above can substitute only for the sweetness of sugar, not its physical bulk. When bulk is important, for example in chewing gums, candies, ice cream, baked goods, and fruit spreads, other types of sugar substitutes, such as sugar alcohols (polyols), may be used. Polyols usually replace sugar on a one-to-one basis (that is, one ounce of polyol substitutes for one ounce of sugar). Since some polyols are not as sweet as sugar, a low-calorie sugar substitute may also be included in the product to provide additional sweetness. Generally, in naming the sugar alcohol, `-oseis changed into `—itol' or `-it' .

Some common sugar alcohols:

- Glycol (2-carbon)

- Glycerol (3-carbon)

- Erythritol (4-carbon)

- Threitol (4-carbon)

- Arabitol (5-carbon)

- Xylitol (5-carbon)

- Ribitol (5-carbon)

- Mannitol (6-carbon)

- Sorbitol (6-carbon)

- Dulcitol (6-carbon)

- Iditol (6-carbon)

- Isomalt (12-carbon)

- Maltitol (12-carbon)

- Lactitol (12-carbon)

- Polyglycitol

The general characteristics of sugar alcohols are nonfermentability, a moistening effect, the maintenance of quality and heat resistance. The Maillard reaction does not occur easily, and thus browning and burning of food are prevented. Sugar alcohols have an oxidation suppression effect on protein degeneration, so permitting the retention of freshness in fish and animal meat. Moreover, they help to reduce caloric intake due to their slow digestion and do not stimulate the secretion of insulin. There is also no rapid elevation of the blood glucose level (in application as sweeteners for diabetic patients), they do not increase lipoprotein-lipase activity, thus helping to prevent obesity, and they suppress oxidation of vitamin C. Sugar alcohols have a cool feeling in the mouth and this feeling results from the fact thatdissolution of sugar alcohol in water is an endothermicreaction.

Polyols and other bulk sugar substitutes have three potential advantages over sugar as food ingredients:

• Unlike sugars, they do not promote tooth decay. The bacteria in dental plaque, which produce substantial amounts of decay-promoting acid from sugars and starches, produce little or no acid from polyols. In the United States, FDA allows a health claim on foods made with polyols stating that the food does not promote tooth decay, provided that the food also meets other requirements (such as not containing decay-promoting sugars). Label claims of this type are often found on sugarless chewing gums made with polyols.

• Polyols produce a lower glycemic response (i.e., a lower rise in blood sugar levels after consumption) than most sugars and starches do. Thus, their use may have advantages for people with diabetes.

• Polyols are lower in calories than sugar is — usually by about half, because they are incompletely digested.

Although it helps with calorie reduction, it can also lead to gastrointestinal effects such as looser stools and gas production (flatulence). Hence incomplete digestion, is a mixed blessing. Gastrointestinal effects of polyols increase with the amount consumed, and some people are more sensitive than others to these effects. In the United States, some products containing substantial amounts of polyols are required to carry a label notice stating that -excess consumption may have a laxative effect."

Important benefits of sugar alcohols from a dental perspective include their non- and little fermentability by oral microorganisms (non-cariogenic nature) in human dental plaque and their ability to promote remineralisation of demineralised enamel.

Xylitol:

The sugar corresponding to xylitol (5 carbons) is xylose and is obtained by hydrogenation of xylose obtained during purification of xylan from cottonseed cake and trees, such as oaks and white birches. Xylitol is found in fruits, such as plums and berries, and in vegetables.

The sweetness of xylitol is similar to that of sucrose, but the sweet taste appears and disappears a little faster. The rapid dissolution of xylitol in water results in a cool feeling the mouth. A large portion of ingested xylitol is directly absorbed by the small intestine and subsequently metabolised and the remainder reaches the large intestine where it is fermented by enterobacteria. The available energy value of xylitol, as a nutrition indicator, is 3kcal/g. The results of several epidemiological studies indicate that xylitol is non cariogenic. The results of in vitro studies have also shown that xylitol is not metabolised by mutans streptococci or other microorganisms in the oral cavity. In support of this, it has also been found that plaque pH is not reduced following the intake of xylitol. Moreover, a bacteriostatic effect of xylitol on mutans streptococci has been demonstrated. Results frombiochemical studies suggested that xylitol is transported via the fructose-PTS of S. mutans and the xylitol-5-phosphate created by this pathway is not metabolised.

It was believed that the xylitol-5-phosphate may have undergone eventual dephosphorylation and was perhaps exported at the expense of ribitol-5-phosphate. This is the so-called xylitol futile cycle. Xylitol decreases plaque formation and the long-term intake of xylitol has been reported to reduce the S. mutans level in saliva and plaque. Habitual xylitol consumption by mothers has also been shown to result in a statistically significant reduction of the probability of mother-child transmission of mutans streptococci and dental caries in their children. Adding xylitol to fluoridated dentifrices (10-20%) has a similar effect. However, it should be noted that the caries reduction is not equated with asignificant microbial effect. The exponential reduction in colony forming units (CFUs) has not been shown. Possibly, the modest reduction in mutans streptococci is due to the `fasting' effect of xylitol on oral microorganisms. The adaptation of S.mutans to xylitol has beenrecognised, although the effect of xylitol-insensitive strains of S.mutans on fermentable carbohydrates and on glucan synthesis from sucrose has not been clarified. It is generally assumed that xylitol is non-cariogenic and an extremely effective sweetener in sweets, but itsanticariogenic effect is yet to be supported by evidence based data.

$$
\begin{array}{c}
CH_2OH \\
H \!-\!\!\!-\!\!\!-\!\!\!-\!\!\!|\!\!\!-\!\!\!-\!\!\!-\!\!\!- OH \\
HO \!-\!\!\!-\!\!\!-\!\!\!-\!\!\!|\!\!\!-\!\!\!-\!\!\!-\!\!\!- H \\
H \!-\!\!\!-\!\!\!-\!\!\!-\!\!\!|\!\!\!-\!\!\!-\!\!\!-\!\!\!- OH \\
CH_2OH
\end{array}
$$

Xylitol

Sorbitol:

The sugar corresponding to sorbitol (d-glucitol) (6 carbons) is glucose. There are three types of crystal: alpha, beta and gamma.Sorbitol (fig.8) is the sugar alcohol most frequently added to food, both in solid and in liquid form. Commercially it is sold in aqueous solution because of its hygroscopic properties. It is found naturally in fruits, including apples, pears, and apricots, as well as seaweed and can be obtained by hydrogenation of glucose. The sweetness of sorbitol is 60-70% that of sucrose, and the sweet taste disappears a little faster than that of sucrose. Sorbitol is metabolised in the same manner as xylitol. and the available energy value is 3kcal/g. How mutans streptococci use sorbitol as a carbon source is well understood from a dental perspective. Although acid formation in the bacterial plaque can occur, sorbitol is considered non-cariogenic in nature because of the slow acid production during its metabolism by oral microorganisms. It is often used as a negative control in dental plaque acid production studies. Sorbitol may cause gastric upset in large doses and acts as a laxative because of osmotic transfer of water into the bowel. The food and agricultural organization WHO commission's report on food additives recommends that the intake of sorbitol be limited to 150mg/kg/day.

Erythritol:

The sugar corresponding to erythritol (4 carbons) is an erythriose. Erythritol (fig.9) exists widely in nature, including in lichen, mushrooms, fruits, fermented foods, and the body fluids of mammals. It is also obtained from the fermentation of glucose by yeast. The sweetness of erythritol is 70-80% that of sucrose and is able to mask the bitter taste of sweeteners such as stevioside, so making it comparable to sucrose in this respect. As a result, it is used to improve the taste of this high-intensity sweetener. Erythritol is predominantly absorbed promptly from the small intestine (90% or more), and most of the absorbed sugar is excreted in urine without being metabolised. Therefore, it does not provide an energy source (the energy value is 0kcal/g), nor does it cause diarrhoea.

Erythritol can be classified as a non-cariogenic sweetener, only one study has reported its cariogenicity. In that study, S.11111tC111S PS-14 and S.sobrinits6715 did not attach to glass in the presence of erythritol, indicating that erythritol does not appear to be used by mutans streptococci for the synthesis of water-insoluble glucans. A significantly lower caries score was observed in the rats infected with S. sobrinus 6715 and fed with erythritol.

$$CH_2OH$$
$$|$$
$$HC-OH$$
$$|$$
$$HC-OH$$
$$|$$
$$CH_2OH$$

Erythritol

Maltitol:

Maltitol, also termed reducing maltose, is a disaccharide alcohol of glucose and sorbitol, obtained by thehydrogenation of maltose. The sweetness ofinaltitol is75-80% that of sucrose and its quality of taste resembles that of sucrose. A portion of the ingested maltitol is hydrolysed by maltase in the small intestine, but most reaches the large intestine where it is fermented by enterobacteria. The available energy value is 2kcal/g. Results from several studies have shown that maltitol is non-cariogenic in nature. Evaluation of maltitolin vivo by the pH response of dental plaque using an intra-oral apparatus and in vivo by experimental enamel demineralisation has demonstrated that maltitol does not lower plaque Ph. It has also been demonstrated that 14 strains of oral streptococci, including mutans streptococci, do not utilise maltitol or produce sufficient acid in its presence to demineralise tooth enamel. Furthermore, maltitol does not serve as a substrate for glucosyltransferases of either S. mutans MT8148R or S.mutans 6715 for the synthesis of water-insoluble glucan.

Lactitol:

Lactitol, also termed reducing lactose, is a disaccharide alcohol of galactose and sorbitol obtained by the dehydrogenation of lactose. Its sweetness is 30-40% that of sucrose, and its quality of taste resembles that of sucrose. A proportion of ingested lactitol is hydrolysed by lactase in the small intestine, but most reaches the large intestine where it is fermented by enterob acteii a.

The available energy value of lactitol is 2kcal/g. Several reports have demonstrated that lactitol is non-cariogenic in nature. It is not easily metabolised by acidogenic and polysaccharide-forming oral microorganisms. It has been foundto have extremely lowenamel-demineralising activity. Its enamel demineralising potential has been found to be low invivo, and acid production and dental plaque formation from lactitol in man have been found to be substantially lower than those of sucrose.

Palatinit:

Palatinit is obtained by the dehydrogenation of palatinose. It is virtually an equimolar mixture of glucopyranosyl- 1,6-sorbitol and glucopyranosyl-1,6-mannitol. The sweetness of Palatinit is 45% that of sucrose, and the quality of its sweetness resembles that of sucrose. The majority of ingested palatinit reaches the large intestine, where it is fermented to organic acid by enterobacteria and subsequently absorbed. The available energy value is 2kcal/g. The results of several studies suggest that palatinit is non cariogenic in nature.

SUGAR SUBSTITUTES-ARE THEY SAFE ??

There is some ongoing controversy over whether artificialsweetener usage poses health risks. A study done in 2005by the University of Texas Health Science Center at SanAntonio showed that, rather than promoting weight loss, theuse of diet drinks was a marker for increasing weight gain andobesity. Those who consumed diet soda were more likely togain weight than those who consumed naturally-sweetenedsoda.Animal studies have convincingly proven that artificial sweeteners cause body weight gain. A sweet taste inducesan insulin response, which causes blood sugar to be storedin tissues, but because blood sugar does not increase with artificial sweeteners, there is hypoglycemia and increased foodintake. So in the experiment, after a while, rats given artificialsweetener have steadily increased caloric intake, increasedbody weight, and increased adiposity. Other adverse effects or health hazards will be discussed with individual compounds.

Aspartame

Aspartame was first approved in 1981. It is 200 times sweeter than sugar, with a caloric value similar to sugar (4 kcal/gram). However, since small amounts are used in foods it is considered essentially free of calories. In the mid-1990's, a researcher raised concerns that a rise in brain cancer incidence was linked to aspartame use. However, after intense testing both in animals and humans, there has been no link to aspartame and cancer. There has also been no evidence to support any other side-effects connected to the sweetener.

Saccharin

Saccharin was discovered in 1879, and was considered safe until 1977, when the FDA proposed a ban on the substance due to concerns about rats that developed bladder cancer after receiving high doses. This prompted the agency to put a ban on the sweetener, as well as a warning on labels. Further studies have since shown that the bladder tumors found in the rats were related to a mechanism in rats not found in humans. Due to these findings, it is no longer listed as a potential cancer causing agent, and the requirement for the warning label has been removed.

Acesulfame-K

Acesulfame-K is 200 times sweeter than sugar, with no calories. It was first approved by the FDA in 1988 for specific uses including as a tabletop sweetener. In 1998, it was then approved for use in beverages. In December 2003, it was approved for general uses in foods, but not in meat or poultry. There have not been any health problems, including cancer, associated with the substance despite more than 15 years ofextensive studies. It is not broken down by the body and is eliminated unchanged by the kidneys. Therefore diabetic patients may safely use the product without it affecting their blood glucose levels.

Neotame and Tagatose

The newest of the low-calorie sweeteners, it was approved by the FDA in 2002 as a general purpose sweetener. It is approximately 7000 times sweeter than sugar. Prior to its approval, neotame was subjected to well over 100 scientific studies. These studies found no link to disease and use of theproduct. Tagatose was recently approved in 2001 It is a low calorie sweetener derived from lactose, a carbohydrate found in many dairy products.

Sucralose

Sucralose is the only non-calorie sweetener made from real sugar. To produce the substance, scientists alter the structure of the sugar molecule, making it much sweeter than sugar. Unlike sugar the body does not recognize it as a carbohydreate, so it does not cause dental caries. Also, sucralose cannot be digested, absorbed or metabolized for energy, nor affecting blood glucose levels, thus making it safe for diabetics. It can also be used safely by people with phenlyketouria. In reviewing studies over the past 20 years, it has not been shown to cause cancer, reproductive, or neurological risks to humans.

Polyols

The most common polyols listed on labels are: erythritol, lactitol, mannitol, sorbitol and xylitol. Polyols offer many benefits. They taste like sugar but have fewer calories than sugar. They don't promote tooth decay and produce a low glycemic response. Thus, consumers, especially those with diabetes, may choose to use them. But, consume those foods that contain them in moderation because they may cause a laxative effect similar to prunes or other high fiber foods.

The Food andDrug Administration (FDA) approves new food additives based on reviews of extensive scientific research on safety. Before a new food additive can go on the market, the company that wishes to sell it must petition the FDA for its approval. The petition must provide convincing evidence that the new additive performrns as intended and is safe, where "safe" means a reasonable certainty of no harm under the intended conditions of use. Demonstrating that an additive is safe is the manufacturer's responsibility; FDA just assess the research results and make decisions on the submitted petitions. For additives that are likely to be widelyused, such as sugar substitutes, the necessary research includes extensive studiesin experimental animals, including studies in which high doses of the additive are administered to two species of animals for the greater part of the animals' lifetime. In many instances, studies in human volunteers are also conducted. Most safety studies on prospective food additives are never published in the scientific literature because they do not make an important contribution to scientific knowledge.

The Indian Parliament has recently passed the Food Safety and Standards Act, 2006that overrides all other food related laws. The Act establishes a new national regulatory body, the Food Safety and Standards Authority of India, to develop science based standards for food and to regulate and monitor the manufacture, processing, storage, distribution, sale and import of food so as to ensure the availability of safe and wholesome food for human consumption. This body approves the food additives to be used in various food products.

No artificial sweetener should play a major role in a healthful diet. Even if all of these sweeteners were given the green light for safety tomorrow, they would still fall short when it comes to good nutrition. Like sugar, sugar substitutes, and many of the foods that contain them contribute little or nothing in the way of nutrients, and also take the place of more nutritious foods in the diet. Limit yourself to a couple of servings a day.

PRACTICAL PROBLEMS IN USE OF SUGAR SUBSTITUTES IN PREVENTIVE DENTISTRY

Search for suitable sweetening agent which will satisfy all the characteristics of sugar is going on through years, since then there is no such substitute which will replace sugar in all aspects, but, cariogenic potential can certainly be reduced by using sugar substitutes. Recently, few sugar substitutes are even considered to have antimicrobial property against caries producing microbes in oral cavity.

The dental professionals have the opportunity to provide advice regarding the importance of diet and role of sugars in caries formation. It is unlikely that many patients will voluntarily restrict their sucrose consumption permanently in order to reduce dental caries keeping in view of the human taste preference for sweetness. So it is important that the dentist must be familiarized with the alternatives to sugars and the types of food products that are available with substitute sweetening agents.

Sucrose is an unusually versatile sweetener, useful in many different types of products, so difficulties in the substitution of sucrose arise as sweeteners differ from each other both in their physical and chemical properties. Replacement of sucrose by other sweeteners is not at all simple because they differ from each other both in their physical and chemical properties. As single sweetener is not able to fulfill all the roles of sucrose in the different products, sweetener that can best imitate the role of sucrose in the product in question should be added. For example, Sugar alcohols, xylitol and sorbitol, replace sucrose wellin most of the products; however they are unsuitable for baking.

Chewing gum was suggested as a practical vehicle for caries prevention, because it allows xylitol to stay in the mouth long enough and requires only a low dose. It is also considered as practical vehicle for nicotine in smoking cessation programs or different medicines. But chewing gum is still considered bad manners or even forbidden in some countries (e.g. Singapore) and in some religious communities. Therefore, search for cost-effective and culturally acceptable caries management strategies is important.

ACCEPTANCE OF SUGAR & SUGAR SUBSTITUTES BY THE PUBLIC

Participation is considered as key in success of any public health programmes. Since the public cannot accept what is not offered to it and food processors will not offer what they think not be accepted, the key elements of acceptance are those used by food technologists and market researchers in the design of new products. Similarity to sucrose taste is only one of several factors. The important ones are legality, price, stability, utility in product classes (e.g. low-calorie foods), ease of advertising the advantage of the product to the consumer, and the ability to induce a purchase motivation through advertising, and education of the consumer, who usually has little or no understanding of terms like sugar substitute or artificial sweetener. Availability, taste preference, physiochemical properties and most importantly their cost and public perception are few areas which influence their acceptance by public. Xylitol containing snack foods were generally well accepted by the children, excellent level of participation of both teachers and children has been demonstrated when xyltiol was used in a clinical trial. It is to be noted that introduction of "sugar less" chewinggum increased the total chewing gum market size without decreasing the consumption of sugar gum. Addition of xylitol in subsidized breakfast and lunch programs in elementary schools is suggested.

CONCLUSION

Without any obligations, all the studies stated that sugar substitutes have some sort of advantage in preventing dental caries. Different substitutes have different mechanisms of caries preventive action. Caries reversal with regular use of xylitol was attributed to its remineralisation effect. The use of xylitol chewing gum is today recommended in Finland and many other countries as a "smarthabir. on an individual level." The various sucrose substitutes have different characteristics, which can each be harnessed if used in combination. For example, adding aspartame or stevioside to maltitol and xylitol has been recommended, as has using a combination of palatinose and xylitol. However adding xylitol to fermentable sugars, such as sucrose, should be avoided. Using sucrose substitutes in all sweets would be an effective public health measure, but this is not a realistic option: Instead, we need to consider how to use sucrose substitutes or non-cariogenic sweets to promote oral health. Each of the sucrose substitutes has particular characteristics that should be utilized so that the requirements of specific individuals are met. The prevalence of dental caries in children is declining, but children at high risk of developing dental caries are still an important public health concern. Practical methods of evaluating an individual's dental caries risk have been established and these methods can be applied in general dental clinics and community health centers. The use of non-cariogenic sweets can be recommended by professionals in these clinical settings as an important adjunct to reducing dental caries risk in individuals. To ensure success, a greater variety of sweets is required and new sucrose substitutes of nutritional value should also be developed. Trusted information about different sugar substitutes should be collected from authorized sites and press releases by government organizations. This can avoid misconception about the various food additives such as sugar substitutes.

BIBLIOGRAPHY

1. Stanislaw jerzylec quotes http://thinkexist.corn/quotes/stanislaw jerzy_leciAccessed on Mar 25 2014.

2. Brown RJ, Banate MA, Kristina IR. Artificial Sweeteners: A systematic review of metabolic effects in youth. Int J Pediatr Obes. 2010; 5(4): 305-312.

3. Isokangas P, Tiekso J, Alanen P, Makinen KK. Long-term effect of xylitol chewing gum on dental caries. Community Dent Oral Epidemiol 1989; 17: 200-203.

4. Kandelman D, Gagnon G. A 24-month Clinical study of the Incidence and Progression of dental caries in relation to consumption of chewing gum containing Xylitol in school Preventive Programs. J Dent Res 1990; 69(11): 1771-1775.

5. Wennerholm K, Arends J, Birkhed D. Effect of Xylitol and Sorbitol in Chewing-Gums on Mutans Streptococci, Plaque pH and Mineral loss of Enamel. Caries Res 1994; 28: 48-54.

6. Honkala E, Rimpela A, Karvonen S, Rimpela M. Chewing of Xylitol gum- A well adopted practice among Finnish Adolescents. Caries Res 1996; 30: 34-39.

7. Alanen P, Holsti ML, Pienihakkinen K. Sealants and xylitol chewing gum are equal in caries prevention. Acta0dontolscand2000; 58: 279-284.

8. Jannesson L, Renvert S, Kjellsdotter P. Nabi N, Birkhed D. Effect of a Triclosan-containing toothpaste supplemented with 10% Xylitol on mutans streptococci in saliva and dental plaque. Caries Res 2002; 36: 36-39.

9. Miyasawa H, Iwami Y, Mayanagi H, Takahashi N. Xylitol inhibition of anaerobic acid production by streptococcus mutans at various pH levels. Oral Microbiollmmunol 2003; 18: 215-219.

10. Makinen KK, Saag M, Isotupa KP, Olak J, Nommela R, Soderling E, et.al. Similarity of the effects of erythritol and Xylitol on some risk Factors of Dental Caries. Caries Res 2005; 39: 207-215.

11. Biria M , Malekafzali B, Kamel V. Comparison of the effect of Xylitol g,uni- and Mastic-chewing on the remineralization rate of caries-like lesions. Journal of Dentistry, Tehran University of Medical Sciences, Tehran, Iran 2009: 6(1): 6-10.

12. Suda R, Suzuki T, Takiguchi R, Egawa K, Sano T, Hasegawa K et al. The effect of adding calcium lactate to xylitol chewing gum on remineralization of enamel lesions. Caries Res 2006; 40: 43-46.

13. Thaweboon S, Nakornchai S, Miyake Y, Yanagisawa T, Thaweboon B, Soo-Ampon S, et.al. Remineralization of enamel subsurface lesions by Xylitol Chewing gum containing Funoran and Calcium Hydrogen phosphate. J Trop Med public Health 2009; 40(2): 345-353.

14. Subramaniam R. Antimicrobial activity of Stevioside on Periodontal pathogens and Candida Albicans- An invitro study. J Indian Assoc Public Health Dent2011; 17: 325-329.

15. Jensen ME. Responses of interproximal plaque pH to snack foods and effect of chewing sorbitol-containing gum. J Am Dent Assoc1986;113:262-266.

16. Grenby TH, Phillips A, Mistry M. Studies of the Dental Properties of Lactitol compared with five other bulk sweeteners in vitro. Caries Res 1989;23:315-319. 17. Dawes C, Macpherson LMD.Effect of nine different chewing-gums and lozenges on salivary flow rate and pH. Caries Res 1992; 26: 176-182.

17. Dawes C.Effect of nine different chewing gums and lozenges on salivary flow rate and pH.Caries Res 1992;26:176-182.

18. Scheie AA, Fejerskov 0, Danielsen B. The effects of Xylitol-containing chewing gums on dental plaque and acidogenic potential. J Dent Res1998;77(7):1547-1552.

19. Makinenet.al. Polyol-combinant saliva stimulants: A 4-month pilot study in young adults.Acta0dontol Scand1998:56:90-94.

20. Simons D, Beighton D, Kidd EAM, Collier FI. The effect of Xylitol and Chlorhexidine acetate/Xylitol chewing, gums on plaque accumulation and gingival inflammation. J ClinPeriodontol 1999;26:388-391.

21. Giersten E. Emberland H, Scheie AA. Effects of mouth rinses with xylitol and flouride on dental plaque and saliva. Caries Res1999:33:23-31.

22. Hujoel et.al.The optimum time to initiate habitual xylitol gum-chewing, for obtaining long-term caries prevention. J Dent Res1999: 78(3): 797-803.

23. Honkala S, Honkala E, Tynjala J, Kannas L.Use of xylitol chewing gum among Finnish school children. Acta0dontolScand1999; 57: 306-309.

24. Lam M, Riedy CA, Coldwell SE, Milgrom P, Craig R. Children's acceptance of Xylitol-based foods. Community Dent Oral Epidemio12000;28:97-101.

25. Alanen P, Isokangas P, Gutmann K. Xylitol candies in caries prevention: results of a field study in Estonian children. Community Dent Oral Epidemio12000;28:218-224.

26. Roberts et.al. How xylitol-containing products affect cariogenic bacteria. J Am pent Assoc 2002; 133:435-441.

27. Assev S, Stig 5, Scheie AA. Cariogenic traits in Xylitol-resistant and Xylitol-sensitive mutans streptococci. Oral MicrobiolImmuno12002;17:95-99.

28. Blicks et.al. Effect of Xylitol on mutans streptococci and lactic acid formation in saliva and plaque from adolescents and young adults with fixed orthodontic appliances. Eur J Oral Sci 2004;112:244-248.

29. Holgerson PL, Blicks SC, Sjostrom I, Oberg M, Twetman S. Xylitol concentration in saliva and dental plaque after use of various Xylitol-containing products. Caries Res 2006;40:393-397.

30. Maehara H, Iwami Y, Mayanagi H, Takahashi N. Synergistic Inhibition by combination of Fluoride and Xylitol on Glycolysis by Mutans Streptococci and its biochemical mechanism. Caries Res2005 ;39:521-528.

31. Honkala E, Honkala S, Shyama M, Mutawa SA. Field trial on caries prevention with Xylitol Candies among Disabled School students. Caries Res2006;40:508-513.

32. Holgerson PL, Sjostrom I, Blicks SC,Twetman S. Dental plaque formation and salivary mutans streptococci in school children after use of Xylitol-containing chewing gum. International Journal of Paediatric Dentistry2007;17:79-85.

33. Badet C, Furiga A, Thebaud N. Effect of Xylitol on an In Vitro model of Oral Biofilm. Oral Health Pro, Dent 2008;6:337-341.

34. Blicks SC, Holgerson PL,Twetman S. Effect of xylitol and xylitol-fluoride lozenges on approximal caries development in high-caries-risk children. Int J Clin Pediatr Dent2008:18:170-177.

35. Fontana et.al. Xylitol: Effect on the Acquisition of Cariogenic Species in Infants. Pediatr Dent 2009;31:257-266.

36. Milgrom P, Ly KA, Tut OK, Mancl L, Roberts MC, Briand K, etal. Xylitol pediatric topical oral syrup to prevent Dental Caries. Arch Pediatr Adolesc iyIed. 2009; 163(7):601-607.

37. Thaweboon S, Nakornchai S, Miyake Y, Yanagisawa T, Thaweboon B, Soo-Ampon 5, etal. Remineralization of enamel subsurface lesions by Xylitol Chewing gum containing Funoran and Calcium Hydrogenphosphate. J Trop Med public Health2009: 40(2):345-353.

38. Hildebrandt G, Lee I, Hodges J. Oral mutans streptococci levels following use of a Xylitol mouth rinse: a double-blind, randomized, controlled clinical trial. Spec Care Dentist2010; 30(2):53-58.

39. Paula VA, Modesto A, Santos KRN, Gleiser R. Antimicrobial effects of the combination of Chlorhexidine and Xylitol.Br Dent J. 2010:209(12):19.

40. Nakai Y, Ishihara SC, Kaji M. Xylitol gum and maternal transmission of Mutans Streptococci. J Dent Res 2010; 89(1):56-60.

41. Fernandes FSF, Fernandes JKB, Marques SG, Silva RA. Effect of chlorhexidine gel containing saccharin or aspartame in deaf children highly infected with mutans streptococci. Brat J Oral Sci2011;10(1):7-11.

42. Simoes MR, Modesto A, Santos KRN, Drake D.The effect of 1% Chlorhexidine Varnish and 40% Xylitol Solution on Streptococcus Mutans and plaque accumulation in children. Pediatr Dent2011; 33(7): 484-490.

43. Ramanarayanan S, Mittal S, Hiregoudar M, Basha, S, Manjunath PG, Natraj CG. Antifungal activity of four commercially available intense sweeteners against candida albicans- An In vitrostudy. UnivRes J Dent 2013;3(2):60-63.

44. Schernhammer ES, Bertrand KA, Birmann BM, Sampson L, Willett WC, Feskanich D. Consumption of artificial sweetener- and sugar-containing soda and risk of lymphoma and leukemia in men and women.Am J ClinNutr. 2012;96(6):1419-1428.

45. Azza A.M. AbdElfatah, Inas S. Ghaly and Safaa M. Hanafy. Cytotoxic Effect of Aspartame (Diet Sweet) on the Histological and Genetic Structures of Female Albino Rats and Their Offspring. Pak J BiolSci2012; 15(19): 904-918.

46. Cong W-n, Wang R, Cai H, Daimon CM, Scheibye KM, Vilhelm AB, et al. Long-Term o Artificial Sweetener Acesulfame Potassium Treatment Alters Neurometabolic Functions in C57BL/6J Mice. PLoS ONE 2013; 8(8): e70257.

47. Kashanian S, Khodaei MM, Kheirdoosh F.In vitro DNA binding studies of Aspartame, an artificial sweetener. J PhotochemPhotobiol B2013;120:104-110.

48. Findikli Z, Tiirkoglu S. Determination of the effects of some artificial sweeteners on human peripheral lymphocytes using the comet assay.J. Toxicol. Environ. Health Sci. 2014; 6(8): 147-153.

49. Mandihassan S. A comparative study of the word sugar and of its equivalents in Hindustani as traceable to Chinese.Am J Chin Med. 1981; 9(3):187-92.

50. Mittal S. Demand-Supply Trends and Projections of Food in India. Indian Council for Research on International Economic Relations. Working Paper No. 209,2008.

51. Gupta S. Food Expenditure and Intake in the NSS 66th Round. Economic & Political Weekly 2012; 157(2): 23-26.

52. Fernie WT. Meals Medicinal: With "Herbal Simples" (of Edible Parts) Curative foods from the cook in place of drugs from the chemist. 1905, Sugars and syrups, Part-4.

53. Sognnaes RF. Analysis of wartime reduction of dental caries in European children. Am J Dis Child. 1948; 75: 792-821.

54. Takeuchi M. Epidemiological study on Japanese children before, during and after World War II. Int Dent J1961; 11: 443-57.

55. Marthaler TM. Epidemiological and clinical dental findings in relation to intake of carbohydrates. Caries Research 1967; 1: 222-38.

56. Roshan NM, Sakeenabi B. Practical problems in use of sugar substitutes inpreventive dentistry. IntSocPrev Community Dent. 2011; 1(1): 1-8.

57. Milgrom P, Ly KA, Tut OK, Mancl L, Roberts MC, Briand Ketal. Xylitol pediatric topical oral syrup to prevent Dental Caries. Arch Pediatr Adolesc Med.2009,163(7):601- 607.

58. Shafer. Textbook of Oral pathology, 5th Edition, Elsevier. 2006: 567-658.

59. Zero DT.Sugars-The ArchoCriminal? Caries Res2004; 38: 277-285.

60. Matsukubo T and Takazoe I. Sucrose substitutes and their role in caries prevention. Int Dent J2006; 56(3): 119-130.

61. Meister K. Sugar substitutes and your health. American council on science and health 2006.

62. Christenson JA. Sugar substitutes- Are they safe? The University of Arizona Cooperative extension. AZ 1229.

www.ingramcontent.com/pod-product-compliance
Lightning Source LLC
Chambersburg PA
CBHW022108170526
45157CB00004B/1536